Modern Hybrid Machining and Super Finishing Processes

This book captures the recent breakthroughs in subtractive manufacturing and difficult-to-machine, material-based, modern machining techniques. It illustrates various combinations of hybrid machining and super finishing, and outlines the critical area profile accuracy, high-precision machining, high tolerance, surface quality, chipping, and cracking for converting into new applications.

Modern Hybrid Machining and Super Finishing Processes: Technology and Applications provides scientific and technological insights on subtractive manufacturing routes. It covers a wide range of micromachining parts, electronic components, metrological devices, and biomedical instruments on materials such as titanium, stainless steel, high-strength temperature-resistant alloys, fiber-reinforced composites, and ceramics, refractories, and other difficult-to-machine alloys. The book emphasizes machined surface accuracy and quality of surface, productivity, and automatization. It also covers creating complex, intricate, and complicated shapes for difficult-to-machine materials. The book goes on to offer an investigation on electrochemical discharge machining, abrasive-based nano-finishing, and rotary ultrasonic machining-based parametric combination, as well as discuss the latest trends in hybrid machining combined processes.

This book is a firsthand reference for commercial organizations mimicking modern hybrid machining processes by targeting difficult-to-machine, materials-based applications. By capturing the current trends of today's manufacturing practices, this book becomes a one-stop resource for scholars, manufacturing professionals, engineers, and academic researchers.

Sustainable Manufacturing Technologies: Additive, Subtractive, and Hybrid

Series Editors: Chander Prakash, Sunpreet Singh, Seeram Ramakrishna, and Linda Yongling Wu

This book series offers the reader comprehensive insights of recent research breakthroughs in additive, subtractive, and hybrid technologies while emphasizing their sustainability aspects. Sustainability has become an integral part of all manufacturing enterprises to provide various techno-social pathways toward developing environmental friendly manufacturing practices. It has also been found that numerous manufacturing firms are still reluctant to upgrade their conventional practices to sophisticated sustainable approaches. Therefore this new book series is aimed to provide a globalized platform to share innovative manufacturing mythologies and technologies. The books will encourage the eminent issues of the conventional and non-conventual manufacturing technologies and cover recent innovations.

Manufacturing Engineering and Materials Science
Tools and Applications
Edited by Abhineet Saini, B. S. Pabla, Chander Prakash, Gurmohan Singh, Alokesh Pramanik

Manufacturing Technologies and Production Systems
Principles and Practices
Edited by Abhineet Saini, B. S. Pabla, Chander Prakash, Gurmohan Singh, Alokesh Pramanik

Modern Hybrid Machining and Super Finishing Processes
Technology and Applications
Edited by Ankit Sharma, Amrinder S. Uppal, Bhargav P. Pathri, Atul Babbar, Chander Prakash

For more information on this series, please visit: www.routledge.com/Sustainable-Manufacturing-Technologies-Additive-Subtractive-and-Hybrid/book-series/CRCSMTASH

Modern Hybrid Machining and Super Finishing Processes

Technology and Applications

Edited by Ankit Sharma, Amrinder Singh Uppal, Bhargav Prajwal Pathri, Atul Babbar, and Chander Prakash

CRC Press
Taylor & Francis Group
Boca Raton London New York

CRC Press is an imprint of the
Taylor & Francis Group, an **informa** business

Designed cover image: © Shutterstock

First edition published 2024
by CRC Press
2385 NW Executive Center Drive, Suite 320, Boca Raton FL 33431

and by CRC Press
4 Park Square, Milton Park, Abingdon, Oxon, OX14 4RN

CRC Press is an imprint of Taylor & Francis Group, LLC

ISBN: 978-1-032-35429-3 (hbk)
ISBN: 978-1-032-35662-4 (pbk)
ISBN: 978-1-003-32790-5 (ebk)

DOI: 10.1201/9781003327905

Typeset in Times
by Apex CoVantage, LLC

Contents

About the Editors

Dr. Ankit Sharma holds the position of Associate Director (Research) at the CURIN Department of Chitkara University, Punjab, India. He also heads the Chikara University Publications/publisher for international peer-reviewed Chitkara journals. He received his doctorate in Mechanical Engineering from the Thapar Institute of Engineering and Technology, Punjab, India. He is a passionate researcher with diversified research interests in manufacturing, medical devices, modern machining processes, additive manufacturing, tissue engineering, developing automated machines for health industries, and not limited to that. To date, he has published more than 55 scientific articles in various peer-reviewed reputed top-notch journals, conferences, and books in manufacturing and materials.

Dr. Sharma is also organizing an international conference and serving as the General chair and convenor of the International Conference on Emerging Materials, Smart Manufacturing, & 3D Printing, Computational Intelligence (ICEMSMCI 2023) with publishing partners Springer, Taylor & Francis, Inder Science, AIP Proceedings, and Bentham Science. He has also been the organizing committee member of the International Conference on the Advances in Materials and Manufacturing Technology-2022. He has chaired various international conferences in the domain of advanced manufacturing. He has delivered several seminars and keynote talks on international (USA, China, India) platforms and awarded with best research paper awards.

Dr. Sharma is working on research commercialization and filed more than 30 patents. He also granted 4 international patents in his account along with 18+ granted patents. Dr. Sharma edited/authored 3 books, serving as a series editor of Taylor and Francis books. He is also serving as a guest editor and Managing editor of international peer-reviewed Journals and books.

Mr. Amrinder Singh Uppal is working as the lead mechanical engineer in Merla Wellhead Solutions, Houston, Texas. He is a licensed professional engineer in Texas, USA. He has more than ten years' industrial experience in the designing and manufacturing of high-pressure equipment that is used in the oil and gas industry. He has coauthored national and international publications. He has been reviewing research articles of various peer review.

Dr. Bhargav Prajwal Pathri is currently working as Assistant Professor in the School of Technology, Woxsen University, Hyderabad. He has completed his PhD from Malaviya National Institute of Technology (MNIT), Jaipur. He received his master's in automotive engineering from Coventry University, UK. He received his bachelor's in mechanical engineering from JNTU Hyderabad. His research interests include ceramic machining, tool design, advance manufacturing processes, rapid prototyping, bio-printing, bio-materials, and CAD/CAM. He has published articles in more than 25 different international journals. He is also reviewer for many reputed journals. He worked as a product design faculty at Effat University, Jeddah, Saudi Arabia.

Dr. Atul Babbar is working as Deputy Dean at the Research and Development Department of SGT University, Gurugram. His research contribution includes, but is not limited to, sustainable manufacturing, modern machining, and additive manufacturing. He has authored more than 50 research articles and book chapters in various international/national Web of Science and Scopus journals. He holds the position of series editor and editor of several Taylor & Francis book series and books, respectively. He is also working as a guest editor for several journals. He has been granted numerous national and international patents. He has been reviewing research articles from various peer-reviewed SCI and Scopus-indexed journals.

Dr. Chander Prakash is Pro-Vice Chancellor, Research & Development, Chitkara University, Punjab, India and serving Professor at Department of Mechanical Engineering. He served as Dean Research & Development at SVKM's Narsee Monjee Institute of Management Studies, Deemed-to-be-University, Mumbai, India and Lovely Professional University, India. He is a Ph.D. alumnus of Panjab University, Chandigarh. He is also an Adjunct Professor (Honorary position) at the Institute for Computational Science, Ton Duc Thang University, Vietnam. He is a dedicated teacher who embraces student-centric approaches, providing experiential learning to his students. He is a passionate researcher with diversified research interests—developing materials for Biomedical and Healthcare applications, additive manufacturing, and developing and exploring new cost-effective manufacturing technologies for biomedical industries. He has published over 397 (326 SCI/Scopus) scientific articles in peer-reviewed, reputable, top-grade journals, conferences, and books. Dr. Prakash is a highly cited researcher at the international level, and he has 8010 citations, an H-index of 50, and an i-10 index of 173. He is among India's Top 1% of leading Mechanical and Aerospace Engineering scientists, Research.com. He holds 38 ranks in India and 1590th rank in the world. He also consistently appeared in the top 2% of researchers as per Stanford Study in 2021, 2022, and 2023. Dr. Prakash edited/authored 25 books, serving as a series editor of two books and as guest editor of several peer-reviewed journals.

He is an Editorial board member of *Journal of Magnesium and Alloys* (Elsevier, IF-17.6, SJR 2.4, Q1); *Nanofabrication* (IF-2.9, Q2); *International Journal on Interactive Design and Manufacturing, IJIDeM* (Springer, IF-2.1); *Cogent Engineering* (Taylor & Francis, WOS, IF: 1.9); *High-Temperature Materials and Processes* (De-Gruyter, IF: 1.5), *Journal of Electrochemical Science and Engineering* (WOS, IF: 2.1); "https://www.springer.com/journal/42824/" Materials Circular Economy: Science, Engineering, and Sustainability (Springer). He is working on Research commercialisation and has published 18 patents. His four patents were granted. Dr. Prakash raised over 1.2 million USD in grants from various national and international bodies, including the Ministry of Science & Technology India, UKIERI-DST and SERB Government of India.

Dr. Prakash organized various international conferences and faculty development programs. He was acknowledged with the Research Excellence Award for the best and most highly productive researcher in 2019, 2020, and 2021 by Lovely Professional University, Phagwara, Punjab.

Contributors

Dr. Mohammed Alshinwan is currently working at the Faculty of Information Technology, Applied Science Private University, Amman, Jordan. He received a PhD degree from the Inje University, Gimhae, Republic of Korea, in 2017. He was an Assistant Professor at Amman Arab University, Jordan. Currently, he is an Associate professor at an Applied science private university, in Jordan. His research interests in computer networks, Mobile networks, information security, AI, and optimization methods.

Dr. Atul Babbar is working as Deputy Dean at the Research and Development Department of SGT University, Gurugram. His research contribution includes, but is not limited to, sustainable manufacturing, modern machining, and additive manufacturing. He has authored more than 50 research articles and book chapters in various international/national Web of Science and Scopus journals. He holds the position of series editor and editor of several Taylor & Francis book series and books, respectively. He is also working as a guest editor for several journals. He has been granted numerous national and international patents. He has been reviewing research articles from various peer-reviewed SCI and Scopus-indexed journals.

Mr. Dhaval Jaydev Kumar Desai currently working as a Senior Piping Engineer at Worley ECR, Texas, 77584, USA. His area of interest is in the domains of oil and gas, refining, petrochemicals, polyethylene, renewable energy, hydrocarbon and advanced manufacturing.

Dr. Aneesh Goyal is an Assistant professor in the Department of Mechanical Engineering, Chandigarh Group of College, Landran, Mohali, Punjab India.

Dr. Harish Kumar Garg is working as an Assistant Professor in the Mechanical Engineering Department. Earlier he served in the Mechanical Engineering Department at Chandigarh Engineering College, Landran, and Chandigarh University. He has completed his PhD from SLIET Longowal. He has an M. Tech from NIT Jalandhar and a B. Tech from SBSCET, Ferozepur. He has more than 18 years of teaching experience. His area of interest is Hybrid Composite materials, non-conventional machining, and Micro and nano machining processes. He is a member of the board of studies at DAV University, Jalandhar. He is a lifetime member of ISTE. He is a faculty incharge of the Robotic Club in the university and has guided students to participate in technical events at various renowned institutes like IITs, PEC, NIT, and CU and has won top positions.

Dr. Satadru Kashyap is currently working as an Assistant professor in the Department of Mechanical Engineering at Tezpur University, Napam, Sonitpur, Assam (India). His research interest is in advanced manufacturing, material science, polymer composites, and ceramics. He is having more than ten years of academic, research, and industrial experience. He has published several articles in peer-reviewed journals.

Dr. Manoj Kumar is working as an assistant professor in the Department of Mechanical Engineering, at Chandigarh University, Mohali, Punjab India. He is an experienced researcher with expertise in manufacturing, surface engineering, and materials processing. Dedicated to advancing knowledge and fostering a passion for learning among students. Skilled in conducting research, optimizing processes, and publishing impactful findings. Adept at utilizing various instrumentation and software tools for surface characterization and analysis. Strong track record of delivering lectures, mentoring students, and contributing to departmental activities.

Mr. Mohit Kumar is a PhD Scholar & currently MR at Auxein Medical Private Limited Sonipat Haryana. His research interests includes Additive Manufacturing, Surface Engineering etc. He has published many research articles of SCI & SCIE journal and having more than 9 years' experience in research & industry.

Dr. Mohit Kumar is an Assistant Professor at the Department of Mechanical Engineering, University Institute of Engineering, Gharaun, Mohali. He has secured a gold medal for his academic excellence in M.Tech. He has published over 16 peer-reviewed papers and book chapters (Research article – 14, Book Chapter – 2) in reputed international/national journals and books. His publications include high-quality journals and their respective categories as Polymer Composites, Journal of Materials Engineering and Performance, Applied Composite Materials, Fibers and Polymers, Journal of Mechanical Science and Technology, etc. Additionally, he is the editorial review board member of two international journals. His research activities have focused on mechanical characterizations, synthesis of polymers, aging studies of FRP composites, Tissue Engineering, and 3D printed bio-composites of waste management.

Dr. Rajinder Kumar is working as an Assistant Professor in the Department of Mechanical Engineering, I.K.G Punjab Technical University, Kapurthala, Punjab, India.

Mr. Rakesh Kumar is a PhD Research Scholar in the Department of Mechanical Engineering at Chandigarh University, Punjab, India, and also worked as a Assistant Professor in the Department of Mechanical Engineering at Chandigarh Group of Colleges, Landran, Mohali, Punjab (India). He is currently working as a research Engineer in the department of Regulatory Affair at Auxein Medical Private Limited Sonipat Haryana. His research interests includes Additive Manufacturing, Biodiesel Production and Surface Engineering and having more than 13 years' experience in teaching & industry.

Dr. Santosh Kumar is working as Associate Professor in the Department of Mechanical Engineering at Chandigarh Group of Colleges, Landran, Mohali, Punjab (India). He has done his M.Tech. with distinction in 2015. He has more than

9 years of teaching & research experience. His research interests includes Surface Engineering, Coating, Rapid Prototyping and Bio fuels on which he has published more than 80 research articles. He has also published many patent, participated in many international conference, organize FDP, write books etc.

Dr. Satish Kumar is an Associate professor in the Department of Mechanical Engineering, Chandigarh Group of College, Landran, Mohali, Punjab India.

Dr. Rahul Mehra is an Associate professor in the Department of Mechanical Engineering, Chandigarh Group of College, Landran, Mohali, Punjab India. More than 12 years of working experience in the Research and Education Field. Published more than 20 international and national research papers. Worked as service Engg. at Honeywell Automation India Ltd for 1 Year.

Dr. Ankit Sharma holds the position of Head of the Chitkara University publications division of the *Research and Innovation Network (CURIN)* Department, Chitkara University, Punjab, India. He has more than ten years of experience in academics, research, consulting, training, and industry. Dr. Sharma is also organizing an international conference-ICEMSMCI 2023 and serving as the general chair and convenor. He has authored numerous national and international publications in SCI, Scopus, and Web of Science–indexed journals. He has filed/published 30 plus international patents. He is also the book series editor of the CRC Press, Taylor & Francis published series entitled "Innovations in Smart Manufacturing for Long-Term Development and Growth". He is the editor of three CRC Press, Taylor & Francis books, with several more in the pipeline. He is the managing and special issue editor of several international journals with databases of SCIE, ESCI, and Scopus-indexed. He has delivered a number of invited seminars and keynote talks on international (USA, China, India) platforms and awarded with best research paper awards.

Prof. Anoop Kumar Singh is a professor of research in the department of the Chitkara University Research and Innovation Network (CURIN) Department, Chitkara University, Punjab, India. He has more than 36 years of experience in academics, research, and the aviation industry. His field of interest includes manufacturing technology, minimum quality l; lubrication, clean technology, roto-dynamic machines, and optimization. He is a fellow of the Institution of Engineers (India) and a life member of the Indian society of Technical Education. He has filed 15 patents, authored 40 research papers, and guided several PhD and M.E. Scholars. He has organized an international conference on Materials and Manufacturing Technology (AMMT-2011).

Dr. Chandandeep Singh is working as an assistant professor in the Mechanical Engineering department at Punjabi University, Punjab, India. His research interest is in the areas of Production and Industrial Engineering, CAD/CAM, Die-Casting Sustainable Green Development, Advanced Manufacturing and Maintenance Techniques. He is a Life member of the International Association of Engineers. iv). Senior Member of UAMAE under The IRED (Institute of Research Engineers and

Doctors. He has published more than 50 research papers and he has been the author/editor of several books with international publishers (Taylor & Francis and Wiley).

Dr. Gagandeep Singh is working as an Assistant Professor in the Department of Mechanical Engineering at Baba Farid College of Engineering and Technology, Bathinda, India.

Mr. Harvinder Singh is a researcher from the Department of Mechanical Engineering, Chandigarh Group of College, Landran, Mohali, Punjab India.

Mr. Jaspreet Singh is working as an assistant professor in the Department of Mechanical Engineering at Chitkara University Institute of Engineering and Technology, Chitkara University, Rajpura, Punjab, India.

Dr. Kamaljeet Singh is currently employed in the Department of Academics, Pune Institute of Business Management, Pune, India. He is an instrumental researcher in the domain of advanced manufacturing and non-conventional machining processes with more than fifteen years of experience.

Dr. Kanwaljit Singh is working as an assistant professor in the department of Mechanical Engineering, Punjab State Aeronautical Engineering Collage, Patiala, Punjab, India. His specialization is in non-conventional machining (ultrasonic machining, chemical assisted ultrasonic machining, electric discharge machining, and water jet machining), and conventional machining (friction stir welding, CNC, and DNC).

Dr. Manpreet Singh is working as an Assistant Professor in the Department of Mechanical Engineering at Baba Farid College of Engineering and Technology, Bathinda, India. He is passionate about bridging academia and industry through innovative research. Proudly contributed to a series of groundbreaking publications in advanced finishing techniques, advanced manufacturing techniques and optimization techniques earning recognition for my research contributions. Dedicated to fostering an intellectually stimulating environment for students, guiding them towards academic excellence. Committed to innovation, with several patent applications showcasing my drive to make a positive impact on society.

Dr. Navdeep Singh is an assistant professor in the Department of Mechanical Engineering, at Sant Baba Bhag Singh University, India.

Dr. Naveen Mani Tripathi is an assistant professor at Assam Energy Institute, Sivasagar, Assam (A center of Rajiv Gandhi Institute of Petroleum Technology), India. He received his PhD from Ben-Gurion University of the Negev, Beer-Sheva, Israel. He has vast international exposure in the domain of Powder and granules rheology, advanced manufacturing, materials, and pipelines. He has published several research papers in reputed journals.

Preface

The book entitled *Modern Hybrid Machining and Super Finishing Processes: Technology and Applications* aims to present comprehensive and most recent breakthroughs in the multidisciplinary area of subtractive manufacturing and difficult-to-machine, material-based, modern machining techniques. The focused theme illustrates various combinations of hybrid machining and super finishing in order to suit the best complex difficult-to-machine, materials-based applications for providing superior finished surface quality and exceptional productivity. Along with this, it will outline the critical areas, exemplified as profile accuracy, high-precision machining, the effect of minimum quantity lubrication (MQL), surface texture, high tolerance, surface quality, chipping, and cracking for converting into new applications.

By reading this book, buyers gain access to practical guidelines and real-world case studies that can directly enhance their knowledge and expertise in hybrid machining and super finishing processes and advancements in electrochemical discharge machining, rotary ultrasonic machining, abrasive-based nano-finishing processes, electric discharge machining, hybrid machining, and advanced finishing processes. Since this book will capture the aforementioned trends of today's modern hybrid machining practices, hence, we are highly confident that this contribution will benefit all the readers in different aspects.

1 A Brief Review on Electrochemical Discharge Machining Process

Satadru Kashyap

1.1 INTRODUCTION

With ever-growing improvements in technology, applications in the micro and nanoscale has risen in varied booming domains, such as biomedical, MEMS, microfluidics, etc. (Judy, 2001; Haeberle & Zengerle, 2007; Dario et al., 2000; Sharma et al., 2020a). However, catering to these industries has become a huge challenge, as they demand high quality and intricacy with optimal use of resources. In this regard, non-conventional machining techniques have played a paramount role in producing machined components at the micro scale. Non-conventional machining techniques can be broadly classified based on the use of energy principles as follows:

(a) Chemical energy—employed by techniques such as chemical machining technique which is used in the machining of intricate designs on different materials with good surface finish. However, they suffer from disadvantages, such as substandard dimensional accuracy and low productivity rates (Çakir et al., 2007).
(b) Mechanical energy—Ultrasonic machining and abrasive jet machining employs mechanical energy to erode particles from the surface in order to machine brittle and hard materials. However, these processes are unsuitable for machining ductile materials (Cheema et al., 2015; Yuvaraj & Kumar, 2015; Sharma et al., 2022a).
(c) Thermal energy—Techniques such as laser beam machining (LBM), electric discharge machining (EDM), micro electric discharge machining (μ-EDM) and wire cut electric discharge machining (WCEDM) employ thermal energy in order to remove materials from a component. While LBM is efficient in machining metals, ceramics as well as glass, however, the presence of a heat-affected zone hinders its extensive usage in industries (Sharma et al., 2022b; Shirk & Molian, 1998). Additionally, EDM processes can only be employed on conducting materials (Geng et al., 2014).

DOI: 10.1201/9781003327905-1

Further, few studies were reported while machining ductile and brittle materials using various machining techniques which shows a futuristic approach to machining techniques (Akhai & Rana, 2022; Babbar et al., 2019, 2020a, 2020b, 2020c, 2020d, 2021a, 2021b, 2021c, 2021d, 2022; Kalia et al., 2022; Khanduja et al., 2021; Kumar et al., 2021; Parikh et al., 2023; Prakash et al., 2021; Rampal et al., 2021; Rana & Akhai, 2022; Sharma & Jain, 2020; Sharma et al., 2018, 2019a, 2019b, 2019c, 2020, 2020a, 2021, 2021a, 2022, 2022a, 2022b, 2022c, 2023a, 2023b; Singh et al., 2021, 2022, 2023).

Thus, there was an imminent requirement for a machining technique that could machine at the micro level with sound dimensional accuracy without consideration of material properties, such as hardness, strength, ductility or conductivity. Electrochemical discharge machining (ECDM) is a hybrid non-conventional machining process that involves the principles of electric discharge machining (EDM) and electrochemical machining (ECM) (Allesu et al., 1992). Developed in 1968 by Kura Fuji, it is used for machining/micromachining both conductive and non-conductive materials such as glass and ceramics irrespective of their properties, such as ductility, hardness, etc. Initially, this process was employed in drilling of glass and ceramics (Kurafuji & Suda, 1968; Sarkar et al., 2006). Subsequently, it was extended to stainless steels and composites (Tandon et al., 1990; Khairy & Mcgeough, 1990) and was employed in micro fabrication applications, such as deep drilling, micro dies, micro profiling and micro grinding (Khairy & Mcgeough, 1990; Jain & Chak, 2000; Furutani & Maeda, 2008; Schopf et al., 2001; Peng & Liao, 2004). However, this process also has its limitations, such as low accuracy and low aspect ratios. In order to overcome these, the basic ECDM process has been modified with the inclusion of external energies which has yield superior consequences.

1.2 ELECTROCHEMICAL DISCHARGE MACHINING (ECDM)

ECDM process has been diagrammatically represented in Figure 1.1, and this involves a DC source providing current between a cathode (tool electrode) and an anode (additional electrode) both immersed in an electrolyte. The tool electrode is stationed directly above the materials to be machined. Due to the formation of an electrolytic cell, bubbles of hydrogen and oxygen gas emerges at the two electrodes as a result of electrolysis. However, above a critical voltage, the amount of hydrogen emerging at the tool surpasses the generation rate of gas bubbles at the surface of the electrolyte. Due to the accumulation of hydrogen gas bubbles near the tool electrode and, subsequently, coalescing of smaller bubbles to form larger gas bubbles, a film of hydrogen gas is developed around the tool (Bhattacharyya et al., 1999). This hydrogen film serves as an insulator around the tool electrode which ceases the current flow to the tool electrode, thereby generating soaring electric field in excess of 10 V/mm. Presence of high-current density and electric field across the dielectric hydrogen film generates an arc discharge (Wuthrich et al., 2005a). During this arc discharge, huge bombardment of electrons occurs on the surface of the workpiece from the tool side which raises the temperature of the workpiece at the locations of bombardment, resulting in the melting and subsequent removal of material (Kulkarni et al., 2002).

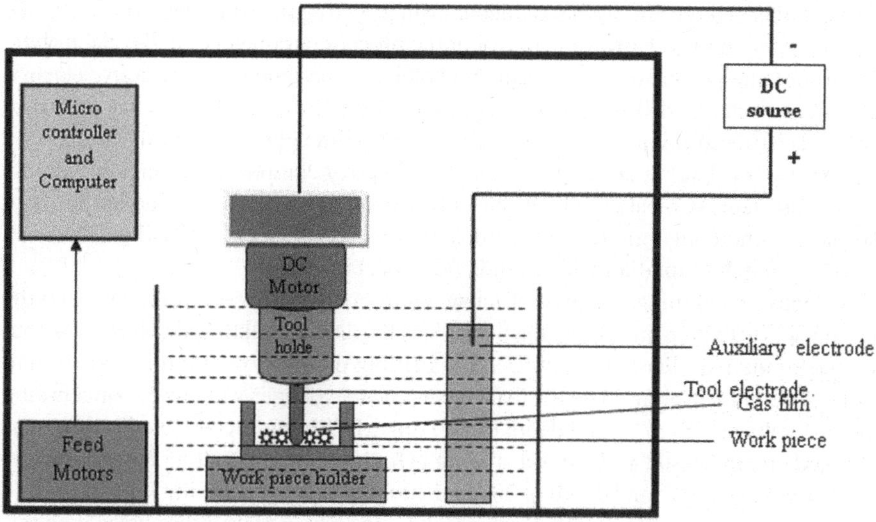

FIGURE 1.1 Schematic view of the electrochemical discharge machining process.

The mechanism of arc discharge has been elaborated by numerous researchers in different manners. Basak and Ghosh revealed the existence of soaring current densities across thin conducting bridges in the hydrogen film near the tool electrode. Presence of this high current density causes a boiling effect in the bridges which leads to arc discharge (Basak & Ghosh, 1996). In another study, researchers hypothesized that each gas bubble was a valve, and bursting of these bubbles under high current densities lead to arc discharge (Jain et al., 1999) showed that the shape of arc discharge was cylindrical as revealed through the markings on the workpiece surface (Behroozfar & Razfar, 2016a). Arc discharge raises the electrolyte temperature which then improves the surface finish of the workpiece via chemical etching of the surface (Yang et al., 2001). Overall, the arc discharge mechanism in ECDM is a complex process which has not been completely elaborated yet and may be considered to be a combination of all previously mentioned.

ECDM process has been very successful in micro fabrication of hard and brittle materials irrespective of their conductivity. Many modified variations of ECDM process as shown in Figure 1.2 has been put forward in different applications before, such as turning, milling, drilling, die-sinking and dressing. These are discussed in the sections later.

1.2.1 Electrochemical Discharge Drilling (ECDD)

ECDM process was earlier used in micro drilling of through and blind holes of conductive materials, such as chromium, low alloy steels, nimonic alloys, titanium, etc. The necessity of drilling is derived from the requirement of micro holes with high aspect ratio in hard, thick and thin materials. Holes fabricated with ECDM process yield high surface finish and dimensional accuracy. Controlled tool movement in the vertical axis and the arc discharge mechanism affecting the workpiece surface make the process complex. However, control of the process parameters via the electrolyte, electrodes and power source (current and voltage) is paramount for effective electrochemical discharge drilling (ECDD) operation. ECDD process being a versatile process was earlier used in the micro drilling of soda-lime glass (Maillard et al., 2007), silicon wafers (Paul et al., 2014), e-glass fiber-epoxy composite (Manna & Narang, 2012) and steels (Coteaţă et al., 2008). It has also been used in developing polished surfaces without any micro cracks, silica micro-scale devices and drill holes of 450 μm deep and 300 μm dia in 30 seconds (Mousa et al., 2009).

Effective machining demands efficient control of the process parameters, and the parameters affecting micro drilling the most is the drilling depth and voltage involved in machining (Maillard et al., 2007). The drilling diameter required is divided into three different zones based on the drilling depth and voltage: (1) Zone A—machining done at voltage (28–37V) and depth of 100 μm. Machining is conducted only in discharge domain. Result yielded is a smooth cylindrical hole. (2) Zone B—machining done at voltage (30V) and depth of 200 and 300 μm. Micromachining is conducted in between the hydrodynamic and discharge domains which yield micro holes with serrated outlines profiles. (3) Zone C—machining done at voltages in excess of 30V and depths in excess of 100 μm. Electrolyte supply at tool is hindered, and as such, speed of drilling depends on depth of drilling and not on the voltage (Maillard et al., 2007). Additionally, liquid soap is added to the electrolyte as a surfactant in order to reduce the film thickness. Addition of surfactant reduces the differences in energy in the intermittent discharges which yield consistent results (Laio et al., 2013). Moreover, superior machining rates are achieved with tool materials having higher thermal conductivity in the discharge domains, but the same tool electrode material yields lower machining rates in the hydrodynamic regime.

Thermally high conductive tool electrodes transfer more thermal energy at the tool surface rather than at the workpiece area. This leads to higher machining rates in the discharge domain than the hydrodynamic domain (Laio et al., 2013; Mousa et al., 2009). Better material removal rates are achieved with rougher tool electrode surface, as it facilitates better wettability and coalescing of the film of gas, thereby providing stability in machining and hole size (Yang et al., 2010). Additionally, spherical-tipped and curved tool surface facilitates better flow of electrolyte in the tool region which decreases the contact area between the two electrodes, thereby facilitating higher machining rates. It has been observed that the machining time reduced by 83%, while the hole diameter reduced by 65% upon using spherical-tipped tool surface as opposed to cylindrical tool used conventionally (Yang et al., 2011). It has also been observed that tool materials such as tungsten carbide and steel perform better than brass due to their inherent wear-resistant behavior (Behroozfar & Razfar, 2016b).

1.2.2 Electrochemical Discharge Turning (ECDT)

ECDM process has also been employed in machining of continuously rotating cylindrical workpieces as in case of turning. Hence, the terminology for this type of machining has been coined as electrochemical discharge turning (ECDT).

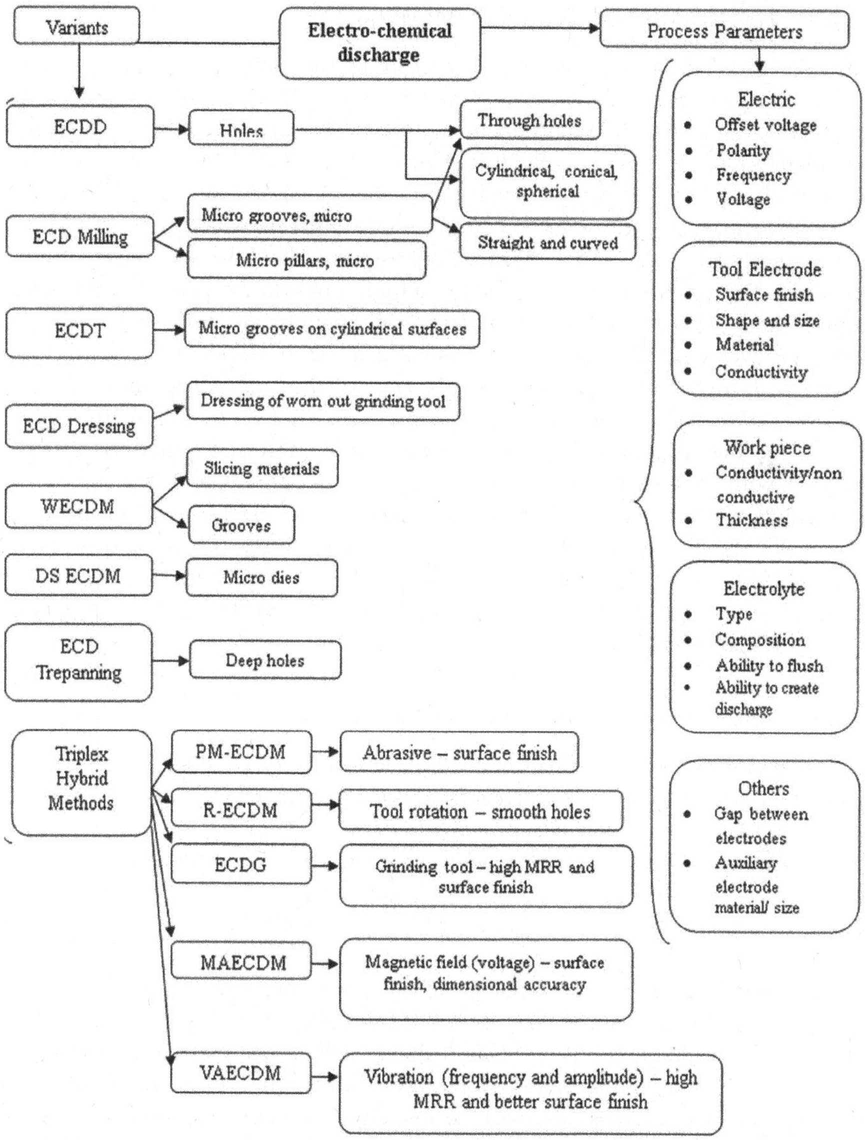

FIGURE 1.2 Electrochemical discharge machining with its different variants and process parameters.

The rotating workpiece immersed in the electrolytic bath facilitates fresh electrolyte being supplied at the tool-workpiece gap. Moreover, it has been observed that optimum rotation of the workpiece being an important parameter would assist in producing narrow, deep grooves with sharp edges on the workpiece surface. However, higher rotation speeds would hinder proper machining due to the inability of gas film formation the workpiece surface (Furutani & Maeda, 2008).

1.2.3 Electrochemical Discharge Milling

In order to fabricate very intricate structure of the micro scale, electrochemical discharge milling has been employed successfully, especially on quartz and glass substrates and while manufacturing micro channels, grooves and texturing their surface (Paul & Hiremath, 2014; Nguyen et al., 2015; Changjian et al., 2012; Ziki et al., 2012). A rotating cylindrical wheel which acts as the tool electrode (cathode) traverses a predefined path in order to mill the workpiece. Machining parameters such as rates of tool travel and rotation affect the machining process the most. For machining micro grooves, a high tool rotation is used as it hinders replacement of electrolyte at the tool electrode, thereby ensuring sharp edges on these grooves with smaller width. Shallow grooves of the micro scale are produced with higher width at high travel speeds of the tool (Zheng et al., 2007a). However, if a deeper micro groove is required, then the process of peeling off workpiece material in a layer-by-layer manner is preferred which yields deep grooves with good surface finish. This is due to the fact that at higher depths, the electrolyte can easily flush of the debris material upon machining if the material removal is done via layer-by-layer manner (Zheng et al., 2007a). It was earlier reported that the results obtained by electrochemical discharge milling for creation of grooves, channels, pillars, etc., of the micro scale has been quite successful (Caoa et al., 2009).

1.2.4 Electrochemical Discharge Dressing

Dressing is the process of repairing a worn-out and dirty grinding tool in order to restore its cutting ability. Electrochemical discharge dressing was coined due to the ability of ECDM process in successfully dressing grinding tools in the micro scale (Schöpf et al., 2001). Herein, the worn-out grinding wheel becomes the cathode immersed in the electrolyte, and another secondary electrode is used as the anode to create the electrical discharge. Energy delivered from the discharge process assists in removing the dirty debris from the surface of the grinding tool by dissociating the metallic bonds on the wheel. Removal of the debris yields newer grains from the grinding tool. Role played by the electrolyte is paramount in this process, as it acts as a cooling medium, debris removal/flushing medium and dielectric fluid. It was observed that the grinding tools dressed through this process require only 50% of the normal grinding forces. This was due to morphology of the grinding tool surface (surface roughness) created through the efficient machining process (Sanjay & Rao, 2008).

1.2.5 Wire Electrochemical Discharge Machining (WECDM)

WECDM process involves a wire as the tool electrode (cathode) for accurate metal removal in the micro scale (Bhuyan & Yadava, 2014). Two types of techniques, *viz.* weight loading technique and reciprocating technique, were put forward in order to ensure efficient contact between the cathode and anode (workpiece) for proper discharge. In case of weight loading technique, the tool electrode (wire) and the job to be machined is held near each other by the aid of mechanical force which hindered the removal of debris from the discharge site. The debris material then assists in generating auxiliary discharges which yields dimensionally inaccurate grooves with bad surface finish. However, in case of reciprocating technique of WECDM, the workpiece is moved to and fro in order to facilitate the flushing of the debris. This yields accurate groove profile with higher surface finish (Yang et al., 2006). SiC abrasive particles are mixed with the electrolyte in order to remove any insulating coating on the tool electrode (wire) which would otherwise enhance the critical voltage. Additionally, these SiC particles act as finishing tools, thereby removing any cracks of the micro level from the machined surfaces (Yang et al., 2006). Slit depth is an important parameter for assessing the efficiency of the process, and its highest ceiling is fixed at 2,000 μm. This slit depth is affected greatly by the feed rate which can be used as high as 350 μm/min. If it is more than 350 μm/min, the gas film at the tool electrode is intermittently broken down, resulting in reduced slit depth (Kuo et al., 2013). Earlier, a new methodology of flow of electrolyte was proposed, whereby a titrated electrolyte is supplied at the discharge site in the form of droplets which reduces the release of toxic fumes during the machining process, thereby reducing pollution (Kuo et al., 2013). WECDM is a low-cost process which has been used in machining a variety of materials irrespective of their electrical conductivity such as quartz (Kuo et al., 2013), ceramics (Tsuchya et al., 1985), glass (Bhuyan & Yadava, 2014) and composites (Jain et al., 1991; Liu et al., 2009).

1.2.6 Die-Sinking Electrochemical Discharge Machining (DS-ECDM)

ECDM process can also be used in manufacturing of shallow dies with better material removal rates than individual EDM and ECM processes (Khairy & Mcgeough, 1990; Panda & Yadava, 2011). Additionally, the DS-ECDM process provides dimensional accuracy similar to EDM and better than ECM process (Khairy & Mcgeough, 1990). A hollow tool made of bronze (3.7 mm inner diameter and 9.4 mm outer diameter) is used with a feed rate of 3–18 mm/min and 120 g/l electrolyte ($NaClO_3$) with a pulsed DC voltage (20–30V) in order to achieve high sinking results (Khairy & Mcgeough, 1990).

1.2.7 Electrochemical Discharge Trepanning

Trepanning is the method to creating deep holes by moving the tool electrode in an orbital motion achieved by offsetting the tool axis from the spindle axis. ECDM process is used efficiently to produce deep holes in brittle and hard materials in a low-cost way. Earlier, deep holes were created in quartz (2.35 mm) and alumina (1.35 mm) using this process (Jain & Chak, 2000). A modified abrasive particle reinforced tool fed with a spring system is used earlier by Chak and Venkateswara Rao (2008) in place of the conventional gravity fed tool in order to achieve better surface finish and depth of hole. The high frequency discharge achieved by the use of abrasives imparts higher material removal rates as well. Additionally, it was also observed that a pulsed DC voltage would assist in better machining efficiency (Sanjay et al., 2007).

1.3 ECDM-BASED TRIPLEX HYBRID METHODS

Further developments and modifications in the ECDM process has led to the development of triplex hybrid methods whereby different energy sources are integrated with ECDM constitution, thus developing a third energy source for discharge. On the basis of this third energy source, the hybrid techniques are further classified into primary and secondary techniques. Generation of energy via rotation of the tool or modification of the electrolyte through the addition of powders consist of primary techniques. While energy discharge for material removal obtained via mechanical and magnetic forces consists of secondary hybrid processes. These triplex hybrid processes can be further classified into the following sections:

1.3.1 Powder Mixed ECDM (PM-ECDM)

Powder mixed ECDM is a triplex hybrid technique whereby abrasive particles are mixed with the electrolyte as the third energy source for material removal process. Abrasive particles such as graphite are used in this process which reduces the effect of energy discharge on the workpiece surface, thus creating a superior surface finish. The reason for reduction in the impact of energy discharge due to the presence of conducting graphite particles may be two folded: (1) development of an intense electric field due to the presence of conductive abrasive particles at the machining area produces a stable energy discharge, and (2) constant movement of abrasive particles results in a continuous flow of charges in between the electrodes. It was previously observed that the surface finish improved significantly with this process, upon addition of 10 mm graphite particles to NaOH electrolyte (Han et al., 2007).

1.3.2 Rotary Electrochemical Discharge Machining (R-ECDM)

In this process, the third energy is achieved from the rotation of the tool electrode which facilitates drilling straight and smooth holes with a small entrance diameter and no cracks. This is due to the fact that as the tool rotation, energy is spread uniformly over the machining area [51–72]. However, care must be taken in order to not go beyond a critical tool rotation, as it creates huge electrolytic turbulence which yields low machining rates due to an unstable discharge. It was observed that in case of drilling pyrex glass with a 200 μm tool diameter, two distinct domains were present—(1) from 500–1,500 rpm tool rotation, the hole entrance diameter gradually reduces with increasing tool rotation and (2) above 1,500 rpm, the hole entrance increases due to reduced sparking discharge and long machining time (Zheng et al., 2007b). Parameter such as the gap distance in between the outer periphery of the tool and the inner periphery of the hole is significant in obtaining deeper holes with little or no taper. Hence, tools are being designed with the facility of reducing or increasing the gap distance. The higher the gap, the smaller is the discharge effect on the hole periphery walls, thus creating micro holes that are deep with reduced taper (Zheng et al., 2007b). Moreover, superior surface finish is achieved through the usage of shorter pulses. Previously, holes with aspect ratio of 11:1 were drilled on borosilicate glass with a tungsten carbide constructed tool electrode (Jui et al., 2013). Additionally, this process was successfully used in drilling micro-scale holes in pyrex glass (Zheng et al., 2007b), borosilicate glass (Jui et al., 2013) and different types of steel (Coteata et al., 2011; Huang et al., 2014).

1.3.3 Electrochemical Discharge Grinding (ECDG)

In this process, the third energy source employs machining via mechanical abrasive action in addition to electric discharge machining and electrochemical dissolution. The tool electrode consists of embedded abrasive grains with metallic bonding in between which serve as a grinding tool which aids in the machining process (El-Hofy, 2005). The metallic bonding network being conductive assists in spark production (Sanjay & Rao, 2008). Feed rate of the tool electrode is paramount, as it allows better contact between the tool (abrasive gains) and the workpiece. A minimum amount of gap between the electrodes and higher surface area of contact (for higher abrasion) will create a thin film of gas with higher sharp edges. This will in turn yield higher material removal with better surface finish. This technique has been previously used efficiently in the machining of cylinders (centerless grinding) and deep holes of the micro scale (Schöpf et al., 2001), machining composites (Wen, 2009), glass and ceramics (Jain et al., 2002).

1.3.4 Vibration-Assisted Electrochemical Discharge Machining (VAECDM)

In case of VAECDM, vibration is applied to either of the electrodes (tool or workpiece) or the electrolyte in order to regularize a continuous and uniform supply of electrolyte at the electrodes (both the tool tip and the job surface) (Wuthrich et al., 2006b; Han et al., 2009; Rusli & Furutani, 2012). With this process, it was possible to drill 300 mm deep holes of the micro scale by applying 0–30 Hz vibrations at the tool. Applying 1.7 MHz frequency vibrations to the electrolyte would produce continuous and uniform discharge at the tool electrode which is significant in achieving higher depths while drilling micro hole with reduced taper (Han et al., 2009). Additionally, the material removal rate increases by two folds on the application of vibrations at the tool electrode (Wuthrich et al., 2006b). In addition to the frequency of the vibrations, the amplitude plays a significant role in the machining process. It was observed that amplitudes <2 mm produces pulsed discharges generating higher material removal than conventional ECDM process, thus, assisting in creating deeper holes. However, amplitudes in the range of 2–3.5 mm yields reduced material removal with better surface finish due to the generation of denser and wider pulsed discharge (Rusli & Furutani, 2012).

1.3.5 Magnetic Field–Assisted Electrochemical Discharge Machining (MAECDM)

In this process, a magnetic field developed via a magnetic system installed as a magnetic tool chuck which holds the tool electrode. Due to the magnetic field developed at the tool electrode, a magneto hydrodynamic convection (MHD) occurs that increases the electrolytic circulation. Enhanced circulation of the electrolyte hinders stable formation of hydrogen gas bubbles, thus breaking them. Hence, a higher voltage is required to maintain a stable gas film at the tool electrode. This higher voltage aids in improving the machining rates and its efficiency. Electrolyte circulation prevents weakening of the gas film which ensures a stable discharge, thereby improving the dimensional accuracy especially in creating deep holes (Liu et al., 2013; Cheng et al., 2010a).

1.4 EFFECTS OF PROCESS PARAMETERS

Hybrid non-traditional machining techniques such as ECDM is an intricate process, the efficiency of which is affected by a host of different parameters, such as electrode gap, polarity, current, voltage, duty cycle (machining parameters), gas film, type of electrode and its properties and tool electrode material and its geometry as shown in Figure 1.2. Hence, the effects of these parameters are paramount for machining efficiency.

1.4.1 Electrolyte Properties

The main role of electrolyte in ECDM process is to etch the workpiece surface chemically and form a stable gas film. A variety of electrolytes are used in ECDM process, viz. NaCl, $NaClO_3$, KOH, H_2SO_4, KCl and even water (Cao et al., 2009; Yang et al., 2001). Additionally, electrolyte properties, viz. viscosity, concentration, conductivity and temperature also affect the machining parameters significantly. It was observed that electrolytes with alkaline nature such as KOH and NaOH aided in improving the material removal rate in ECDM process when compared with acidic (H_2SO_4 or HCl unable to machine glass) or neutral electrolytes (NaCl and KCl results in lower machining rates) (Yang et al., 2001). In another study, it was revealed that addition of soap solution (surfactant) reduces the gas film at the tool electrode which in turn improves wettability and decreases the spark discharge; this improves the surface finish (Jiang et al., 2015). Additionally, electrolytes with graphite powder hinder cracks and enhance surface quality due to the abrasive action provided by graphite powder (Han et al., 2007).

Among the alkaline electrolytes with similar concentration, KOH seemed to improve the material removal rate more when compared with NaOH due to its low viscosity which assists in enhancing the flushing rates at the tool electrode (Cao et al., 2009; Yang et al., 2001). Acidic electrolytes are hazardous in nature due to the fumes and gases exuded by them. Mineral water can be used as an eco-friendly electrolyte and serves good purpose due to the presence of ions such as Na^+, OH-, and H_3O^+ in it. Thus, properties of electrolytes such as concentration, temperature, conductivity, type and viscosity are paramount in defining the ECDM process.

1.4.2 Tool Electrode Properties

Properties of tool electrodes such as thermal conductivity, tool material, shape of the tool electrode and availability of abrasive grains at the tool surface are paramount in enabling an efficient machining process. Researchers have used a host of materials such as tungsten carbide, high-speed steel, high-carbon steel, copper and stainless steel as the tool electrode. However, it has been revealed that tungsten carbide and stainless steel yielded the best results due its high hardness, wear resistance, melting point and low specific heat capacity. High thermally conductive materials used as the tool electrode resulted in high machining efficiency in the discharge domain, while the efficiency is limited in the hydrodynamic regime as a result of the drag force created by the molten material in the upward direction (Mousa et al., 2009). Another significant parameter affecting the machining characteristics is the tool electrode geometry. Over the years, tool electrodes of different shapes and size were used in machining, viz. conical, cylindrical, tubular, spherical, abrasive coated, textured on the surface, etc. Earlier, it was revealed that needle-shaped tool yielded deeper holes when compared with cylindrical shaped tools due to high spark discharge at a concentrated place (Wuthrich et al., 2006a). Spherical electrodes reduced the contact area between the electrodes, thereby enabling fresh supply of electrolyte at the tool site which resulted in higher machining rates and stable gas film (Yang et al., 2011). Tool electrodes with abrasive coating improved the machining rates due to high spark discharge and extra abrasive action of the tool electrode. Additionally, an insulated coating of abrasive particles in the inner periphery of a hollow electrode resulted in enhanced surface finish and dimensional accuracy (Chak & Venkateswara Rao, 2008). Earlier, usage of a flat side tool improved the electrolyte circulation at higher machining depths due to a stable gas film and spark discharge. Hence, deeper holes of the micro scale were possible by using flat side tool (Zheng et al., 2007b). Moreover, tubular electrodes with higher inner diameter enabled better flushing action by the electrolyte at the site of machining. This resulted in higher machining rates (Zhang et al., 2016).

1.4.3 Gas Film Properties

A gas film is formed at the tool electrode due the evaporation of the electrolyte and coalescence of hydrogen bubbles at the tool surface. This gas film is instrumental in defining the surface quality of the machined area, as it defines the surface finish and formation of overcuts and delamination on the surface (Wuthrich & Ziki, 2009). A stable gas film is important for efficient machining process and parameters, such as electrolyte temperature, gas production rate, availability of fresh electrolyte, current density and bubble removal rate (Vogt & Thonstad, 2003; Kelogg, 1950). It was also revealed earlier that tool electrode geometry such as conical tool or tool with insulated side assists in stable gas film formation resulting in better machining process (Wuthrich et al., 2006b). Higher electrolyte circulation with a flat side tool also aids in gas film stability, thereby improving machining rates (Cheng et al., 2010a). Additionally, stability of the gas film is also dependant on electrolyte and tool electrode temperature.

1.5 FUTURE PROSPECTIVE OF ECDM PROCESS

In this article, investigations carried out in the field of electrochemical discharge machining (ECDM) have been discussed in context to the varied modifications and processes developed in this area. Additionally, effects of the various process parameters in regard to these various modified forms of ECDM processes have been presented. Although extensive research has been researched out in this domain, there are still potential for numerous developments and modifications to improve the efficiency and versatility of the machining process.

- It has been revealed through the earlier survey that the ECDM process finds extensive application in the drilling of micro holes and micro channels with reduced taper and lower aspect ratios using ECDD and electrochemical discharge milling processes. If the different types of textures on the surface can be generated in these micro channels through this process, then it may find significant applications in the fields of micro-fluidics.
- The properties of the electrolyte are paramount in the performance of ECDM and its related machining processes. Hence, there is a tremendous potential in developing an efficient, environmentally suitable and cheap electrolyte which will increase the efficiency of the machining technique.
- Industrial applications always ask for low-cost but high-performance processes which would produce high productivity and quality (surface integrity and dimensional accuracy) through these machining processes (i.e. ECD and its variants). Hence, developmental studies in tweaking the process parameters and further modifications in the existing process in order to achieve higher efficiency may be conducted.
- Researcher must review the feasibility of creating a system which facilitates the fabrication of all the micromachining features in the same system. This would increase the commercial viability of the machining process in the fields of semiconductors and MEMS.
- Extensive research may also be conducted toward machining of new generation of materials, such as composites and super alloys with complex textures.

1.6 CONCLUSION

A brief survey on the research works conducted in the field of electrochemical discharge machining and its different variants has been presented. The main conclusions made from this study are as follows:

- ECDM process and its variants are significant non-traditional machining techniques which can be applied for machining both ductile and brittle materials and both conducting and non-conducting materials. It has been widely employed in machining a host of different materials, such as glasses, ceramics, steels, metals and composites.
- The main advantage of this process and its variants is the production of machined features (holes, channels, profiles, etc.) of the micro scale on hard-to-machine materials with superior MRR and surface finish.
- Variants of ECDM process such as R-ECDM, VAECDM and MAECDM are employed to achieve even better MRR and drilling performance during the creation of deeper holes with superior surface finish.

- Other variants of ECDM process, *viz.* electrochemical discharge dressing, is used in cleaning worn-out grinding tools of the micro scale; electrochemical discharge turning is used in machining cylindrical features; wire electrochemical discharge machining is used in segmenting materials; and electrochemical discharge grinding is employed in creating holes on cylindrical parts.

REFERENCES

Akhai, S., & Rana, M. (2022). Taguchi-based grey relational analysis of abrasive water jet machining of Al-6061. *Mater. Today Proc.* 65, 3165–3169. http://doi.org/10.1016/j.matpr.2022.05.361

Allesu, K., Ghosh, A., & Muju, M.K. (1992). Preliminary qualitative approach of a proposed mechanism of material removal in electrical machining of glass. *Eur. J. Mech. Eng.* 36, 202–207.

Babbar, A., Jain, V., & Gupta, D. (2019). Thermogenesis mitigation using ultrasonic actuation during bone grinding: A hybrid approach using CEM43° C and Arrhenius model. *J. Braz. Soc. Mech. Sci. Eng.* 41, 1–14. https://doi.org/10.1007/s40430-019-1913-6

Babbar, A., Jain, V., Gupta, D., & Agrawal, D. (2021a). Histological evaluation of thermal damage to Osteocytes: A comparative study of conventional and ultrasonic-assisted bone grinding. *Med. Eng. Phys.* 90, 1–8. https://doi.org/10.1016/j.medengphy.2021.01.009

Babbar, A., Jain, V., Gupta, D., & Agrawal, D. (2021c). Finite element simulation and integration of CEM43° C and Arrhenius Models for ultrasonic-assisted skull bone grinding: A thermal dose model. *Med. Eng. Phys.* 90, 9–22. https://doi.org/10.1016/j.medengphy.2021.01.008

Babbar, A., Jain, V., Gupta, D., Agrawal, D., Prakash, C., Singh, S., & Bogdan-Chudy, M. (2021b). Experimental analysis of wear and multi-shape burr loading during neurosurgical bone grinding. *J. Mater. Res. Technol.* 12, 15–28. https://doi.org/10.1016/j.jmrt.2021.02.060

Babbar, A., Jain, V., Gupta, D., Prakash, C., & Agrawal, D. (2022). Potential application of CEM43° C and Arrhenius model in neurosurgical bone grinding. In *Numerical Modelling and Optimization in Advanced Manufacturing Processes* (pp. 145–158). Cham: Springer International Publishing. https://doi.org/10.1007/978-3-031-04301-7_9

Babbar, A., Jain, V., Gupta, D., Prakash, C., & Sharma, A. (2020b). Fabrication and machining methods of composites for aerospace applications. In *Characterization, Testing, Measurement, and Metrology* (1st ed., pp. 109–124). Boca Raton, FL: CRC Press.

Babbar, A., Jain, V., Gupta, D., Prakash, C., Singh, S., & Sharma, A. (2020a). 3D bioprinting in pharmaceuticals, medicine, and tissue engineering applications. In *Advanced Manufacturing and Processing Technology* (1st ed., pp. 147–161). Boca Raton, FL: CRC Press.

Babbar, A., Jain, V., Gupta, D., Prakash, C., Singh, S., & Sharma, A. (2020c). Effect of process parameters on cutting forces and osteonecrosis for orthopedic bone drilling applications. In *Characterization, Testing, Measurement, and Metrology* (1st ed., pp. 93–108). Boca Raton, FL: CRC Press.

Babbar, A., Jain, V., Gupta, D., & Sharma, A. (2020d). Fabrication of microchannels using conventional and hybrid machining processes. In *Non-Conventional Hybrid Machining Processes* (1st ed., pp. 37–51). Boca Raton, FL: CRC Press.

Babbar, A., Rai, A., & Sharma, A. (2021d). Latest trend in building construction: Three-dimensional printing. *J. Phys. Conf. Ser.* 1950, 012007.

Basak, I., & Ghosh, A. (1996). Mechanism of spark generation during electrochemical discharge machining: A theoretical model and experimental verification. *J. Mater. Process. Technol.* 62, 46–53.

Behroozfar, A., & Razfar, M.R. (2016a). Experimental and numerical study of material removal in electrochemical discharge machining (ECDM). *Mater. Manuf. Process.* 31, 495–503.

Behroozfar, A., & Razfar, M.R. (2016b). Experimental study of the tool wear during the electrochemical discharge machining. *Mater. Manuf. Process.* 31, 574–580.

Bhattacharyya, B., Doloi, B.N., & Sorkhel, S.K. (1999). Experimental investigations into electrochemical discharge machining (ECDM) of non-conductive ceramic materials. *J. Mater. Process. Technol.* 95, 145–154.

Bhuyan, B.K., & Yadava, V. (2014). Experimental study of traveling wire electrochemical spark machining of borosilicate glass. *Mater. Manuf. Process.* 29, 298–304.

Çakir, O., Yardimeden, A., & Ozben, T. (2007). Chemical machining. *Arch. Mater. Sci. Eng.* 28, 499–502.

Cao, X.D., Kimb, B.H., & Chua, C.N. (2009). Micro-structuring of glass with features less than 100 mm by electrochemical discharge machining. *Precis. Eng.* 33, 459–465.

Chak, S.K., & Venkateswara Rao, P. (2008). The drilling of Al_2O_3 using pulsed DC supply with a rotary abrasive electrode by the electrochemical discharge process. *Int. J. Adv. Manuf. Tech.* 39, 633–641.

Changjian, L., An, G., Meng, L., & Shengyi, Y. (2012). The micro-milling machining of pyrex glass using the electrochemical discharge machining process. *Adv. Mater. Res.* 403, 738–742.

Cheema, M.S., Dvivedi, A., & Sharma, A.K. (2015). Tool wear studies in fabrication of microchannels in ultrasonic micromachining. *Ultrasonics.* 57, 57–64.

Cheng, C.P., Wu, K.L., Mai, C.C., Hsu, Y.S., & Yan, B.H. (2010a). Magnetic field-assisted electrochemical discharge machining. *J. Micromech. Micro Eng.* 20, 7.

Cheng, C.P., Wu, K.L., Mai, C.C., Yang, C.K., Hsu, Y.S., & Yan, B.H. (2010b). Study of gas film quality in electrochemical discharge machining. *Int. J. Adv. Manuf. Tech.* 50, 689–697.

Coteata, M., Schulze, H.P., & Slatineanu, L. (2011). Drilling of difficult-to-cut steel by electrochemical discharge machining. *Mater. Manuf. Process.* 26, 1466–1472.

Coteaţă, M., Slătineanu, L., Dodun, O., & Ciofu, C. (2008). Electro chemical discharge machining of small diameter holes. *Int. J. Mater. Form.* 1, 1327–1330.

Dario, P., Carrozza, M.C., Benvenuto, A., & Menciassi, A. (2000). Micro-systems in biomedical applications. *J. Micromech. Microeng.* 10, 235–244.

El-Hofy, H. (2005). *Advanced Machining Processes.* New York, NY: McGraw-Hill.

Furutani, K., & Maeda, H. (2008). Machining a glass rod with a lathe-type electro-chemical discharge machine. *J. Micromech. Microeng.* 18, 8.

Geng, X., Chi, G., Wang, Y., & Wang, Z. (2014). Study on micro rotating structure using micro wire electrical discharge machining. *Mater. Manuf. Process.* 29, 274–280.

Haeberle, S., & Zengerle, R. (2007). Microfluidic platforms for lab-on-a-chip applications. *Lab Chip.* 7, 1094–1110.

Han, M.S., Min, B.K., & Lee, S.J. (2007). Improvement of surface integrity of electro-chemical discharge machining process using powder-mixed electrolyte. *J. Mater. Process. Technol.* 191, 224–227.

Han, M.S., Min, B.K., & Lee, S.J. (2009). Geometric improvement of electrochemical discharge micro-drilling using an ultrasonic vibrated electrolyte. *J. Micromech. Micro Eng.* 19, 8.

Huang, S.F., Liu, Y., Li, J., Hu, H.X., & Sun, L.Y. (2014). Electrochemical discharge machining micro-hole in stainless steel with tool electrode high-speed rotating. *Mater. Manuf. Process.* 29, 634–637.

Jain, V.K., & Chak, S.K. (2000). Electrochemical spark trepanning of alumina and quartz. *Mach. Sci. Technol.* 4, 277–290.

Jain, V.K., Choudhury, S.K., & Ramesh, K.M. (2002). On the machining of alumina and glass. *Int. J. Mach. Tools Manuf.* 42, 1269–1276.

Jain, V.K., Dixit, P.M., & Pandey, P.M. (1999). On the analysis of the electrochemical spark machining process. *Int. J. Mach. Tools Manuf.* 39, 165–186.

Jain, V.K., Rao, P.S., Choudhury, S.K., & Rajurkar, K.P. (1991). Experimental investigations into travelling wire electrochemical spark machining (TW-ECSM) of composites. *J. Eng. Ind.* 113, 75–84.

Jiang, B., Lan, S., Wil, K., & Ni, J. (2015). Modelling and experimental investigation of gas film in micro- electrochemical discharge machining process. *Int. J. Mach. Tool Manu.* 90, 8–15.

Judy, J.W. (2001). Microelectro mechanical systems (MEMS): Fabrication, design and applications. *Smart Mater. Struct.* 10, 1115–1134.

Jui, S.K., Kamaraj, A.B., & Sundaram, M.M. (2013). High aspect ratio micromachining of glass by electrochemical discharge machining (ECDM). *J. Manuf. Process.* 15, 460–466.

Kalia, G., Sharma, A., & Babbar, A. (2022). Use of three-dimensional printing techniques for developing biodegradable applications: A review investigation. *Mater. Today Proc.* 2022.

Kelogg, H.H. (1950). Anode effect in aqueous electrolysis. *J. Electrochem. Soc.* 97, 133–142.

Khairy, A.B.E., & Mcgeough, J.A. (1990). Die-sinking by electro erosion-dissolution machining. *CIRP Ann. Manuf. Technol.* 39, 191–195.

Khanduja, P., Bhargave, H., Babbar, A., Pundir, P., & Sharma, A. (2021). Development of two-dimensional plotter using programmable logic controller and human machine interface. *J. Phys. Conf. Ser.* 1950, 012012.

Kulkarni, A., Sharan, R., & Lal, G.K. (2002). An experimental study of discharge mechanism in electrochemical discharge machining. *Int. J. Mach. Tools Manuf.* 42, 1121–1127.

Kumar, S., Sudhakar, R.P., Goyal, D., & Sehgal, S. (2021). Process modelling for machining Inconel 825 using cryogenically treated carbide insert. *Met. Powder Rep.* 76, 66–74. http://doi.org/10.1016/j.mprp.2020.06.001

Kuo, K.Y., Wu, K.L., Yang, C.K., & Yan, B.H. (2013). Wire electrochemical discharge machining (WECDM) of quartz glass with titrated electrolyte flow. *Int. J. Mach. Tools Manuf.* 72, 50–57.

Kurafuji, H., & Suda, K. (1968). Electrical discharge drilling of glass. *Ann. CIRP.* 16, 415–419.

Laio, Y.S., Wu, L.C., & Peng, W.Y. (2013). A study to improve drilling quality of electrochemical discharge machining (ECDM) process. *Proc. CIRP.* 6.

Liu, J.W., Yue, T.M., & Guo, Z.N. (2009). Wire electrochemical discharge machining of Al_2O_3 particle reinforced Aluminium alloy 6061. *Mater. Manuf. Process.* 24, 446–453.

Liu, J.W., Yue, T.M., & Guo, Z.N. (2013). Grinding-aided electrochemical discharge machining of particulate reinforced metal matrix composites. *Int. J. Adv. Manuf. Technol.* 68, 2349–2357.

Maillard, P., Despont, B., Bleuler, H., & Ẅuthrich, R. (2007). Geometrical characterization of micro-holes drilled in glass by gravity-feed with spark assisted chemical engraving (SACE). *J. Micromech. Micro Eng.* 17, 1343–1349.

Manna, A., & Narang, V. (2012). A study on micro machining of e-glass fibre—epoxy composite by ECSM process. *Int. J. Adv. Manuf. Technol.* 61, 1191–1197.

Mousa, M., Allagui, A., Ng, H.D., & Wüthrich, R. (2009). The effect of thermal conductivity of the tool electrode in spark-assisted chemical engraving gravity-feed micro drilling. *J. Micromech. Micro Eng.* 19, 7.

Nguyen, K.H., Lee, P.N., & Kim, B.H. (2015). Experimental investigation of ECDM for fabricating micro structures of quartz. *Int. J. Precis. Eng. Manuf.* 16, 5–12.

Panda, M.C., & Yadava, V. (2011). Intelligent modelling and multi objective optimization of die sinking electrochemical spark machining process. *Mater. Manuf. Process.* 27(1), 10–25.

Parikh, P., Sharma, A., Trivedi, R., Roy, D., & Joshi, K. (2023). Performance evaluation of an indigenously-designed high-performance dynamic feeding robotic structure using advanced additive manufacturing technology, machine learning and robot kinematics. *Int. J. Interact. Des. Manuf. (IJIDeM)*, 1–29.

Paul, L., & Hiremath, S.S. (2014). Characterisation of micro channels in electrochemical discharge machining process. *Appl. Mech. Mater.* 490, 238–242.

Paul, L., Hiremath, S.S., & Ranganayakulu, J. (2014). Experimental investigation and parametric analysis of electro chemical discharge machining. *Int. J. Manuf. Technol. Manag.* 28 57–79.

Peng, W.Y., & Liao, Y.S. (2004). Study of electrochemical discharge machining technology for slicing non-conductive brittle materials. *J. Mater. Process. Technol.* 149, 363–369.

Prakash, C., Kumar, V., Mistri, A., Sharma, A., Uppal, A.S., Babbar, A., & Pathri, B.P. (2021). Investigation of functionally graded adherents on failure of socket joint of FRP composite tubes. *Materials.* 14(2021), 6365.

Rampal, R., Goyal, T., Goyal, D., Mittal, M., Dang, R.K., & Bahl, S. (2021). Magneto-rheological abrasive finishing (MAF) of soft material using abrasives. *Mater. Today Proc.* 45, 51140–5121. http://doi.org/10.1016/j.matpr.2021.01.629

Rana, M., & Akhai, S. (2022). Multi-objective optimization of Abrasive water jet Machining parameters for Inconel 625 alloy using TGRA. *Mater. Today Proc.* 65, 3205–3210. http://doi.org/10.1016/j.matpr.2022.05.374

Rusli, M., & Furutani, K. (2012). Performance of micro-hole drilling by ultrasonic-assisted electro-chemical discharge machining. *Adv. Mater. Res.* 445, 865–870.

Sanjay, K.C., Chak, P., & Rao, V. (2007). Trepanning of Al_2O_3 by electro-chemical discharge machining (ECDM) process using abrasive electrode with pulsed DC supply. *Int. J. Mach. Tools Manuf.* 47, 2061–2070.

Sanjay, K.C., & Rao, P.V. (2008). The drilling of Al_2O_3 using a pulsed DC supply with a rotary abrasive electrode by the electrochemical discharge process. *Int. J. Adv. Manuf. Technol.* 39, 633–641.

Sarkar, B.R., Doloi, B., & Bhattacharyya, B. (2006). Parametric analysis on electrochemical discharge machining of silicon nitride ceramics. *Int. J. Adv. Manuf. Technol.* 28, 873–881.

Schöpf, M., Beltram, I., Boccadoro, M., & Kramer, D. (2001). ECDM (Electrochemical discharge machining) a new method for trueing and dressing of metal-bonded diamond grinding tools. *CIRP Ann. Manuf. Technol.* 50, 125–128.

Sharma, A., Babbar, A., Tian, Y., Pathri, B.P., Gupta, M., & Singh, R. (2022c). Machining of Ceramic Materials: A state of the art review. *Int. J. Interact. Des. Manuf. (IJIDeM)*, 1–21. https://doi.org/10.1007/s12008-022-01016-7

Sharma, A., Fidan, I., Huseynov, O., Ali, M.A., Alkunte, S., Rajeshirke, M., Gupta, A., Hasanov, S., Tantawi, K., Yasa, E., Yilmaz, O., Loy, J., & Popov, V. (2023a). Recent inventions in additive manufacturing: Holistic review. *Inventions.* 8(4), 103. https://doi.org/10.3390/inventions8040103

Sharma, A., Grover, V., Babbar, A., & Rani, R. (2020). A trending nonconventional hybrid finishing/machining process. In *Non-Conventional Hybrid Machining Processes* (1st ed., pp. 79–93). Boca Raton, FL: CRC Press.

Sharma, A., & Jain, V. (2020). Experimental investigation of cutting temperature during drilling of float glass specimen. In *IOP Conference Series: Materials Science and Engineering* (Vol. 715, No. 1, p. 012050). IOP Publishing. https://doi.org/10.1088/1757-899X/715/1/012050

Sharma, A., Jain, V., & Gupta, D. (2018). Characterization of chipping and tool wear during drilling of float glass using rotary ultrasonic machining. *Measurement.* 128, 254–263.

Sharma, A., Jain, V., & Gupta, D. (2019a). Comparative analysis of chipping mechanics of float glass during rotary ultrasonic drilling and conventional drilling: For multi-shaped tools. *Mach. Sci. Technol.* 23, 547–568.

Sharma, A., Jain, V., & Gupta, D. (2019b). Multi-shaped tool wear study during rotary ultrasonic drilling and conventional drilling for amorphous solid. *Proc. Inst. Mech. Eng. Part. E. J. Process Mech. Eng.* 233, 551–560.

Sharma, A., Jain, V., & Gupta, D. (2019c). Tool wear analysis while creating blind holes on float glass using conventional drilling: A multi-shaped tools study. In *Advances in Manufacturing Processes: Select Proceedings of ICEMMM 2018* (pp. 175–183). Singapore: Springer. https://doi.org/10.1007/978-981-13-1724-8_17

Sharma, A., Jain, V., & Gupta, D. (2021). Effect of pre and post tempering on hole quality of float glass specimen: For rotary ultrasonic and conventional drilling. *Silicon.* 13, 2029–2039.

Sharma, A., Jain, V., & Gupta, D. (2022a). Mathematical approach on chipping volume estimation generated during rotary ultrasonic drilling for float glass. *Proc. Natl. Acad. Sci. India Sect. A Phys. Sci.* 92, 285–291.

Sharma, A., Jain, V., Gupta, D., & Babbar, A. (2020a). A review study on miniaturization. In *Advanced Manufacturing and Processing Technology* (1st ed., pp. 111–131). Boca Raton, FL: CRC Press.

Sharma, A., Kalsia, M., Uppal, A.S., Babbar, A., & Dhawan, V. (2022b). Machining of hard and brittle materials: A comprehensive review. *Mater. Today Proc.* 50, 1048–1052.

Sharma, A., Kumar, V., Babbar, A., Dhawan, V., & Kotecha, K. (2021a). Experimental investigation and optimization of electric discharge machining process parameters using grey-fuzzy-based hybrid techniques. *Materials* 14, 5820.

Sharma, A., Sandhu, H.S., Goyal, D., Goyal, T., Jarial, S., & Sharda, A. (2023b). Sustainable development in cold gas dynamic spray coating process for biomedical applications: Challenges and future perspective review. *Int. J. Interact. Des. Manuf. (IJIDeM)*, 1–17. https://doi.org/10.1007/s12008-023-01474-7

Shirk, M.D., & Molian, P.A. (1998). A review of ultrashort pulsed laser ablation of materials. *J. Laser Appl.* 10, 18–28.

Singh, B.P., Singh, J., Bhayana, M., & Goyal, D. (2021). Experimental investigation of machining nimonic-80A alloy on wire EDM using response surface methodology. *Met. Powder Rep.* 76, 9–17. http://doi.org/10.1016/j.mprp.2020.12.001.

Singh, B.P., Singh, J., Bhayana, M., Singh, K., & Singh, R. (2022). Experimental examination of the machining characteristics of Nimonic 80-A alloy on wire EDM. *Mater. Today Proc.* 69, 291–296. http://doi.org/10.1016/j.matpr.2022.08.537

Singh, J., Singh, C., & Singh, K. (2023). Rotary ultrasonic machining of advance materials: A review. *Mater. Today Proc.* http://doi.org/10.1016/j.matpr.2023.01.159

Tandon, S., Jain, V.K., Kumar, P., & Rajurkar, K.P. (1990). Investigations into machining of composites. *Precis. Eng.* 12, 227–238.

Tsuchya, H., Inoue, T., & Miyazaki, M. (1985). Wire electrochemical discharge machining of glass and ceramics. *Bull. Japan Soc. Precis. Eng.* 19, 73–74.

Vogt, H., & Thonstad, J. (2003). Review of the cause of anode effects in aluminium electrolysis. *Aluminium.* 79, 98–102.

Wen, L.J. (2009). *Grinding-aided Electrochemical Discharge Machining of Metal Matrix Composites.* Dissertation, The Hong Kong Polytechnic University, Hong Kong.

Wuthrich, R., Comninellis, C.H., & Bleuler, H. (2005a). Bubble evolution on vertical electrodes under extreme current densities. *Electrochim. Acta.* 50, 242–246.

Wuthrich, R., Despont, B., Maillard, P., & Bleuler, H. (2006a). Improving the material removal rate in spark-assisted chemical engraving (SACE) gravity-feed micro-hole drilling by tool vibration. *J. Micromech. Micro Eng.* 16, N28–N31.

Wuthrich, R., Fujisaki, K., Couthy, P., Hof, L.A., & Bleuler., H. (2005b). Spark assisted chemical engraving (SACE) in microfactory. *J. Micromech. Micro Eng.* 15, 276–280.

Wuthrich, R., Spaelter, U., Wu, Y., & Bleuler, H. (2006b). A systematic characterization method for gravity-feed micro-hole drilling in glass with spark assisted chemical engraving (SACE). *J. Micromech. Microeng.* 16, 1891–1896.

Wuthrich, R., & Ziki, J.D.A. (2009). *Micromachining Using Electrochemical Discharge Phenomenon* (2nd ed.). Oxford: Elsevier.

Yang, C.K., Cheng, C.P., Mai, C.C., & Hwa, Y.B. (2010). Effect of surface roughness of tool electrode materials in ECDM performance. *Int. J. Mach. Tools Manuf.* 50, 1088–1096.

Yang, C.K., Wu, K.L., Hung, J.C., Lee, S.M., Lin, J.C., & Yan, B.H. (2011). Enhancement of ECDM efficiency and accuracy by spherical tool electrode. *Int. J. Mach. Tool. Manu.* 51, 528–535.

Yang, C.T., Ho, S.S., & Yan, B.H. (2001). Micro hole machining of borosilicate glass through electrochemical discharge machining (ECDM). *Key Eng. Mater.* 196, 149–166.

Yang, C.T., Song, S.L., Yan, B.H., & Huang, F.Y. (2006). Improving machining performance of wire electrochemical discharge machining by adding SiC abrasive to electrolyte. *Int. J. Mach. Tools Manuf.* 46, 2044–2050.

Yuvaraj, N., & Kumar, M.P. (2015). Multiresponse optimization of abrasive water jet cutting process parameters using TOPSIS approach. *Mater. Manuf. Process.* 30, 882–889.

Zhang, Y., Xu, Z., Zhu, Y., & Zhu, D. (2016). Effect of tube-electrode inner structure on machining performance in tube electrode high-speed electrochemical discharge drilling. *J. Mater. Process. Tech.* 231, 38–49.

Zheng, Z.P., Cheng, W.H., Huang, F.Y., & Yan, B.H. (2007a). 3D micro structuring of pyrex glass using the electrochemical discharge machining process. *J. Micromech. Micro Eng.* 17, 960–966.

Zheng, Z.P., Su, H.C., Huang, F.Y., & Yan, B.H. (2007b). The tool geometrical shape and pulse off time of pulse voltage effects in a pyrex glass electrochemical discharge micro drilling process. *J. Micromech. Micro Eng.* 17, 265–272.

Ziki, J.D.A., Didar, T.F., & Wüthrich, R. (2012). Micro texturing channel surfaces on glass with spark assisted chemical engraving. *Int. J. Mach. Tools Manuf.* 57, 66–72.

2 A Review of Process Parameters of Rotary Ultrasonic Machining

Jaspreet Singh, Chandandeep Singh, Kanwaljit Singh

2.1 INTRODUCTION OF ADVANCED MACHINING PROCESS

Increasing competition in the market and rising demand for better product performance have inspired the development of a wide variety and high-quality materials, including ceramics, composites, carbides, glasses, diamonds, etc., that have superior features like high hardness and instability, etc. These characteristics provide the material its dominant performance; yet using conventional machining techniques, it is challenging to precisely machine these materials. As a result, the material's quality and several crucial attributes are degraded and large machining costs are generated. In recent years, many machining techniques utilizing various energy resources like mechanical, electrochemical, and thermal have been developed to process these materials (Jain & Jain, 2001; Jain, 2011; Wang et al., 2020a).

USM is one of these cutting-edge technologies and has the ability to work with tough materials (Vinod & Khamba, 2010; Agarwal, 2015; Nath et al., 2012; Bhosale et al., 2014). However, the abrasives in the slurry used for static USM cause some dimensional abnormalities in regards of hole fineness, conicity, out of roundness, and hole over size (Jain et al., 2011; Kataria et al., 2015). As a result, the static USM method's limitations were resolved by the development of the rotary ultrasonic machining method (Ning et al., 2016). A comprehensive approach is rotary ultrasonic machining having collaboration of typical diamond grinding and ultrasonic machining operating principles to produce superior MRR and good surface finish than USM and diamond grinding alone. When both traditional and ultrasonic grinding take place at once in rotary ultrasonic machining, there is a removal of material as microchips in addition to the grinding stroke (Churi, 2010; Khoo et al., 2008; Legge, 1966, 1964). The work material is being fed into continuously while still being vibrated ultrasonically by a spinning hollow drill enhanced with diamond abrasive. The amount of coolant is continuously delivered along the center of the drilling, brushing away fragments, preventing tool congestion, and cooling the cutting zone (Pei et al., 1995; Churi et al., 2010; Jiao et al., 2005). Figure 2.1 shows the essential parts of the RUM, including the ultrasonic transducer, spindle, electric motor, ultrasonic power supply unit, machining zone, coolant tank, transformer, and air compressor, among others. Irrespective of their properties, numerous various

DOI: 10.1201/9781003327905-2

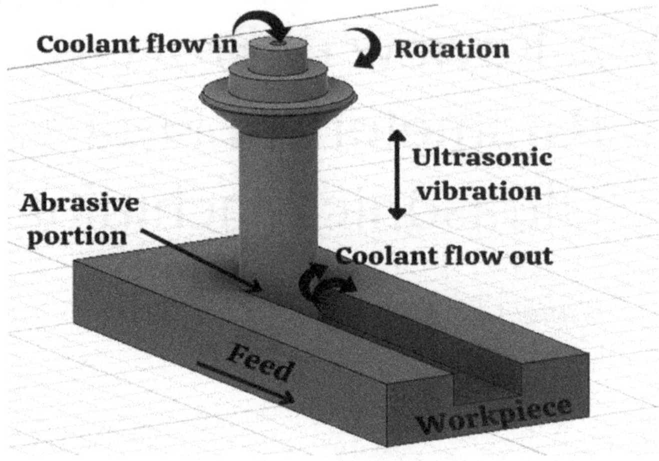

FIGURE 2.1 Diagram of rotary ultrasonic machining.

TABLE 2.1
Classification of Conventional and Non-conventional Parameters

Conventional Process	Non-conventional Process
• Lathe machine	• Electric discharge machining
• Milling machine	• Laser beam machining
• Shaper machine	• Ultrasonic machining
• Slotting machine	• Rotary ultrasonic machining
• Grinding machines	• Chemical machining

types of soft and hard metals are machined using RUM. Several researchers have previously claimed that this mechanism is nonthermal (Churi et al., 2010; Sarwade, 2010; Sharma et al., 2020a). Rotary ultrasonic machining can be classified with various traditional and non-traditional techniques. The development of conventional and non-conventional machining techniques is an outcome of progress in technology. Abrasive jet machining (AJM), water jet machining, electric discharge machining, ultrasonic machining, and rotary ultrasonic machining have great potential in cutting-edge developments (Akhai & Rana, 2022; Babbar et al., 2021a, 2020a, 2020b, 2020c, 2020d; Kalia et al., 2022l; Khanduja et al., 2021; Kumar et al., 2021; Prakash et al., 2021; Rampal et al., 2021; Rana & Akhai, 2022; Sharma et al., 2018b, 2019a, 2019c, 2020; Sharma & Jain, 2020, 2020d, 2021a, 2021b, 2021c; Singh et al., 2021). Further, this study has purely analyzed the necessity of selecting the best parameter combination for the rotary ultrasonic machining method to attain noteworthy outcomes in the form of machining efficiency (Sharma et al., 2022a, 2022b, 2022c, 2023a, 2023b, 2023c, Singh et al., 2022, 2023). The material removal process or the material is removed with the help of machining, and the removed material is in the form of microchips and can be expressed in Table 2.1.

2.1.1 Evaluation of USM to RUSM

Progression from USM to RUM was allegedly created as an upgrade to USM, according to previous studies. In rotary ultrasonic machining, the revolving core drill's surface was filled with diamond abrasives tip as a substitute for the abrasive slurry used in ultrasonic machining. RUSM was traditionally used to drill holes through brittle, hard materials. The USM method, which was first developed in 1927, entailed continually injecting abrasives over the region between the workpiece and tool (Ning et al., 2016). At the UK Atomic Energy Authority in 1964, Mr. Percy Legge developed the RUM technique for the first time. Moreover, Brown et al., were the ones who originally suggested combining drilling and ultrasonic vibrations. In this patent work, drilling facilitated by a very low vibration frequency "on the order of 1 kHz" are solely suggested for boring in wood. In contrast to static USM, P. Legge's initial RUM machine tool had a number of improvements. The many modifications performed included rotating the workpiece while it was being held in a four-jaw chuck and replacing the slurry in the ultrasonic machining the tool material is coated with diamond abrasives. The rotation of the worktable also had some negative effects, such as the device's inability to process larger workpieces and the restriction to only cutting circular holes. As a result of ongoing work at UKAEA, a machine was connected to the rotating-type ultrasonic transducer in Legge's patented technology, which offered a better improvement. This newly developed rotating transducer head allowed for the highly accurate dimensional processing of nonrotating workpieces (Ning et al., 2016; Sharma et al., 2018a). Using a particular kind of tool allowed for the achievement of several process changes. RUM may be used to successfully practice a number of types, including side-to-side milling, surface texture, slot machining, screw threading, and internal-to-external grinding. In order to address USM's shortcomings, RUM had been seen as an improvement. The literature included descriptions of the RUM process in a variety of additional methods. As a hybrid process, that combines USM and traditional diamond grinding as a manufacturing technique (Sharma et al., 2019a). The terms ultrasonic drilling, ultrasonic grinding, ultrasonic vibration grinding, and ultrasonic twist drilling are also occasionally used to describe this process. Workpiece rotation is one of the variations of USM that Komaraiah and Reddy described as part of the term RUM.

2.2 ROTARY ULTRASONIC MACHINING PRINCIPLE

Figure 2.2 shows the working mechanism and the various essential components used in the RUM. Electricity at 50 Hz is changed as 20,000 electric energy in the RUM. Using a piezoelectric transducer, this high-frequency electric power is subsequently transformed into oscillating motion. The spindle and tool are connected, and the tool oscillates in the direction of feed with the assistance of electric motor. Vibration amplitude can be changed using an ultrasonic power source. Electric motors can produce a variety of speed variations for tools (Singh & Singhal, 2016; Pei & Ferreira, 1998). The pressure gauge controls the flow of the cutting fluid, which is fed from the drill's center. The coolant's function is to lower the temperature of the machining

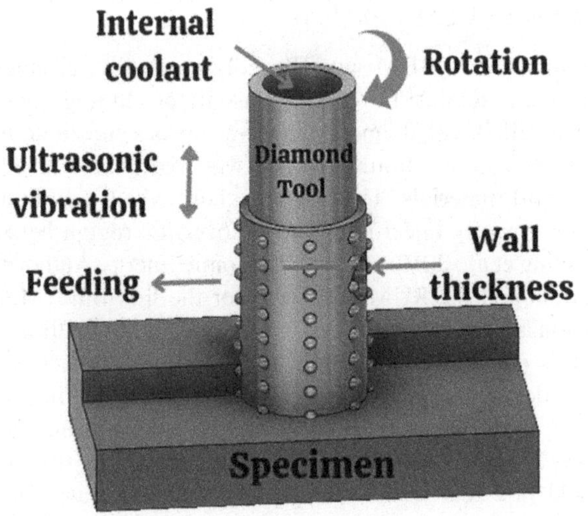

FIGURE 2.2 Illustrate a diagram of RUSM.

zone area (Khoo et al., 2008; Sharma et al., 2019a). The various components are discussed in the given figure later.

2.3 COMPONENTS OF ROTARY ULTRASONIC MACHINING

2.3.1 Ultrasonic Spindle

The machine tool's primary element in RUSM is the ultrasonic spindle, which is made up of a horn, high-frequency generator, transducer, and spindle. With the assistance of a high-frequency generator, electrical energy can be converted into ultrasonic waves of the order of 20 kHz (50–60 Hz) (Sarwade, 2010; Kuo & Tsao, 2012). Electric energy is transformed into high-frequency mechanical vibration using a piezoelectric transducer.

2.3.2 Ultrasonic Power

The tool's vibrational amplitude is controlled by the ultrasonic mechanism of power generation. The amplitude produced during vibration will be getting improved with the increase of ultrasonic power (Cong et al., 2014; Churi et al., 2007). To obtain ultrasonic power ranging from 0% to 100%, RUSM finds its utility more effective than others. Whenever the power supply is turned off, RUM will convert to traditional diamond drilling.

2.3.3 Data Acquisition System

This system can be integrated with the machine tool to measure a variety of reactions, including depth of cut, vibration amplitude, and torque. It consists of an analog-to-digital converter for converting analog signals into digital ones, a dynamometer for measuring torque as well as cutting forces, and a computer program for collecting, plotting, and analyzing the results (Pei et al., 1994; Jiao et al., 2005).

2.3.4 Coolant

A complete system, including a pump, coolant reservoir, filter, pressure controller, indicators, and valves, makes up the cooling device. The hollow tool's center supplies the cutting fluid, which can also be supplied externally via hoses. To give a better finish, cutting fluid is used to remove the extra material and make the surface clean and clear of debris (Pei et al., 1995).

2.4 MECHANISM OF ROTARY ULTRASONIC MACHINING

Following are descriptions of the USM and conventional diamond grinding material removal mechanisms used in the RUM process (illustrated in Figure 2.3) (Ya et al., 2002; Pei et al., 1995; Wang et al., 2020):

- **Hammering**—It occurs when the diamond-based tool and the abrasive impinged on the workpiece. The diamond tool vibrates ultrasonically and crushes the workpiece material with striking and crushing of work material as a hammering action.
- **Abrasive**—Similar to the grinding process, abrasion results from the rotation of the tool concerning the workpiece.
- **Extrusion**—Due to the tool's rotation and longitudinal vibrations working together, extraction was made much easier.

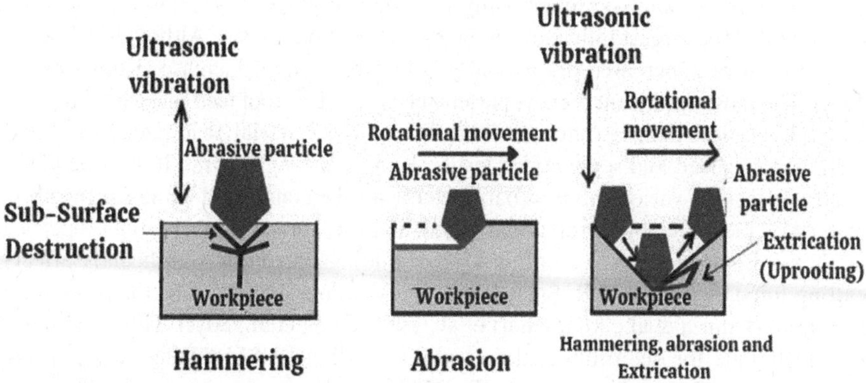

FIGURE 2.3 Shows the methods for removing material in RUSM.

2.5 APPLICATIONS OF RUSM

1. A variety of materials have already been processed using RUM, namely, advanced ceramics, composites, and advanced glasses (Sarwade, 2010).
2. RUM also manufactures advanced glasses like zirconia, which are widely used in the automobile and optical sectors (Kumar et al., 2018; Abdo et al., 2013; Zhang et al., 2014).
3. Stainless steel, which is used in automobiles, medical equipment, and home appliances, has been successfully machined by using this technique (Cong et al., 2010).
4. The application of RUM in orthopedic surgery is by drilling bones without separation (Babbar et al., 2020c).

2.6 PROCESS PARAMETERS AND THE RESPONSE OF RUSM

For the literature review, it should be noted that several studies have been analyzing the findings and comparing the input parameters (such as ultrasonic frequency, rotational speed, concentration, abrasive slurry, grit size, and tool material) to the output responses (such as MRR, SR, TWR, and hole conicity) having a major impact. In this study, we compare the input factors that have a major impact on every output response.

2.6.1 Material Removal Rate

After the literature review, it is to be concluded that MRR is among the most essential factors and is widely used in the production industry for higher material removal rate. While performing the experimentation on rotary ultrasonic machining, it is to be concluded that, as compared to solid tools, it has been found that hollow tools greatly increase the rate of material removal. A hollow tool produces an MRR that is nearly twice as high as a solid tool (Wang et al., 2016). When compared to traditional grinding, the MRR achieved in RUM is increased (Ning et al., 2016; Babbar et al., 2020b). During RUSM of quartz ceramics, the material is extracted primarily through brittle fracture and a small amount of plastic distortion (Singh & Singhal, 2018). In an experiment using BK7 optical glass, it was observed that the feed rate had the largest influence on the operational parameter. Although the material removal rate increased dramatically with even a small increase (Kumar et al., 2018). The most significant process parameter for MRR is tool feed rate (Sindhu et al., 2018). BK7 glass demonstrates a mixed flow of the material at low feed rates and high spindle speed and sonic power levels (Kumar & Singh, 2018). It must be noted that the effects of various process parameters must be considered while cutting alloy material (ultrasonic power, rotational speed, and feed rate) and machining properties while using rotary USM. According to their findings, feed rate significantly affects cutting force, MRR, and SR. Surface roughness values decreased as the power rate increased (Dhuria et al., 2011; Churi et al., 2007; Cong et al., 2011). This is a brand-new technique for determining the amplitude during RUSM (Wang et al., 2009). The influence of many factors, including the cutting tool, vibration amplitude, ultrasonic power, rotational speed, and feed rate, was demonstrated by the results. Only ultrasonic power was determined to be the single most significant factor that had an impact on the vibration's amplitude (Wang et al., 2009).

2.6.2 Surface Roughness

Another factor addressed by the RUSM input variables is the surface roughness of the machined rod and the drilled hole. The SR under the influence of several process factors must be aware, as several RUM research have been carried out. When potassium dihydrogen phosphate (KDP) crystals were machined using rotary ultrasonic technology, they were scrutinized at the surface roughness. Results show that feed rate enhanced surface roughness. When contrasted with a right-angle corner tool, a chamfered corner tool gave a superior finish (Guzzo et al., 2003; Zhang et al., 2014; Babbar et al., 2021a). RUM research was done on optical K9 glass machining (Zhang et al., 2014). The impact of process factors such as ultrasonic power, spindle speed, and feed rate was investigated on the output variables of edge chipping size, cutting power, SR, and power utilization. RUM was found to have fractured the optical K9 glass workpiece while drilling a hole in it using pressurized air as a coolant. The SR of the machined gap facing was significantly affected by raising the feed rate and force of the ultrasonic power; moreover, it is significantly reduced when the spindle was increased. To study the impact of ultrasonic vibration on the surface produced, Pérez et al. (2016) and Babbar et al. (2022) conducted RUM in the surface texturing of CFRP. When the surface with the least vibration frequency was machined together with a tiny input action, a smooth surface was produced.

2.6.3 Tool Wear Rate

The RUM technique results in very little tool wear, particularly when working with brittle materials like CFRP composites and ceramics. For the aim of evaluating tool wear, several investigators have also taken into account subsequent experiments, the tool's linear dimensions have worn, and its weight has decreased. While machining a material of CFRP with rotary ultrasonic machining, it is to be observed that tools wear out a little frequently and had better life spans (Wang et al., 2020). During machining on alumina-based ceramics material, the results show that by using RUM rather than traditional grinding, less tool wear was observed (Gong et al., 2010). A study on the tool life in RUM of CFRP/Ti stacks in terms of the number of holes drilled was done experimentally. Using this procedure, about 250 holes were bored (Cong et al., 2012). As a result of RUM's superior hybrid machining technique, which created less cutting force throughout the operation than cutting force produced by the standard drilling and grinding procedures separately, an improvement in hole quality was seen (Cong et al., 2012, 2014).

2.7 SUMMARY

Table 2.2 provides an overview of how rotary ultrasonic control parameters affect MRR, CF, TWR, SR, and chipping size.

TABLE 2.2
Summary of Previous Rotary Ultrasonic Machining Research Initiatives

Author (Year)	Workpiece Material	Design of Exp. (DOE) Process Parameters (Input)	Machining Characteristics (Output)	Results/Conclusions
Pei and Ferreira (1998)	Ceramic material (magnesia-stabilized zirconia)	Rotary ultrasonic, vibration frequency (50 Hz–20 KHz), rotational speed (3,000 rpm), grit size (270/320), coolant water	Material removal rate (MRR) Static force	RUM scrutinized that as the tool rotation speed increases, the material removal rate also increases so as far the abrasives particles decrease. Further, the increase in removal rate indicates that the static forces increase.
Hocheng and Kuo (2002)	Moulding steel SKD 61	Ultrasonic machining process, frequency (20.69 KHz), amplitude (3.9 μm) Grit size (220), abrasives (aluminium oxide), medium (water) Concentration (40%)	Surface finish Static load	Ultrasonic machining works on moulding steel and copper tool. The polishing is used in ultrasonic machining so that the surface finish is enhanced. Because the static stress on ultrasonic machining has increased, the surface quality must be better.
Choi et al. (2007)	Plane glass	Vibration frequency (20 KHz), low amplitude (2–50 μm), abrasives material (silicon carbide), SiC diameter of the tool (1.5 mm)	Surface roughness (SF) Material removal rate (MRR)	Ultrasonic machining gives rise to the comparison between ordinary ultrasonic machining and material removal rate. While machining plane glass, the MRR and SF of the material must be improved by 200%.
Khoo et al. (2008)	Magnesium (zirconia and alumina)	Vibration frequency (20 kHz)	MRR, TWR, SR	RUM results specify, as the static load increases, the rate of material removal also increases. Moreover, as the surface roughness of the material increases, the amplitude and grit size also increase during machining on magnesium.
Cong et al. (2010)	Stainless steel	Speed (4,000 rpm) Federate (0.02 mm/s) Ultrasonic power (30%)	Surface roughness (SF) Cutting force (CF) Torque	According to RUSM, when cutting force and toque drop, the input parameters, including spindle speed and feed rate, do as well. Surface roughness is improved as spindle speed and feed rate increase.
Liu et al. (2012)	Alumina	Spindle speed (1,500–5,000 RPM)	Cutting force (CF)	Rotary ultrasonic machining results shows the relationship between input parameters (spindle speed, federate, and vibration amplitude) over cutting force, which shows that cutting force increase or decrease depends upon the spindle speed.

Cong et al. (2012)	(CFRP)	Feed rate (0.1–0.7) Ultrasonic frequency (20 KHz), power rate (20–80%), rotational speed (1,000–5,000 rpm)	Tool rotation Power consumption	Rotary ultrasonic machining resulted that the tool rotation decreases as ultrasonic power increases. On the other hand, the consumption of power is higher (i.e. 65%) than that of the rotary machining system.
Abdo et al. (2013)	Zirconia ceramic	Power consumption (50%–70%), depth (.025–.075 mm) Spindle speed (2,000–6,000 r/min), feed rate (50–150 mm/min) Vibration frequency (20–23 KHz)	Surface roughness (SF)	Rotary ultrasonic machining indicates that the value of surface roughness is 0.4295 μm at the spindle speed of 6,000 r/min, feed rate is 50 mm/min, and the depth of cut is 0.025 mm.
Kuo and Tsao (2012)	Non-tempered glass	Rotational speed (5,000 r/min), feed rate (2–6 mm/min), depth (1–3 mm)	Surface roughness (SF)	Rotary ultrasonic–assisted milling of brittle material. The surface roughness is maximum (i.e. 2.56 μm). The feed rate is 0.3 mm/sec and the depth of cut is 3 mm.
Zhang et al. (2014)	Optical K9 glass	Rotary ultrasonic machining, ultrasonic power (0%–30%), spindle speed (2,000–3,000 r/min) Feed rate (0.01–0.02 mm/s)	Cutting force (CF) Edge chipping size Surface roughness (SF)	Rotary ultrasonic machining resulted that an increase in surface roughness on the machined surface tends to decrease in cutting force (CF). Moreover, the power consumption is dependent on the feed rate and spindle speed. It tends to decrease as the spindle speed and federate chipping size increase.
Lv et al. (2013)	Glass BK7	Feed rate (200–600 mm/min), rotational speed (3,000–12,000 r/min) (2–5 μm), frequency (23.1–26.9 KHz)	Sub-surface damage	Rotary ultrasonic machining examined, with the increase of spindle speed, the sub-surface damage will slightly have affected.
Tabatabaei et al. (2013)	Aluminium (65J)	Ultrasonic-assisted machining, natural frequency (228.5 Hz), chip thickness (6 mm), ultrasonic frequency (20 KHz) Stiffness (2.628 N/m)	MRR Surface finish Tool stability	In ultrasonic-assisted machining, the material removed from the workpiece is in the formation of chips. During the machining process, the MRR rate of machining must be improved and the surface roughness is more than that of conventional machining. The tool life and the tool stability are also more than the conventional machining process.
Baek et al. (2013)	Soda-lime glass	Amplitude (20 μm), frequency (20KHz) Tool WC abrasives: (aluminium oxide) Concentration 30 Feed rate (10 μm/s)	Hard wax Without hard wax Surface finish Material removal rate (MRR)	Ultrasonic machining is done on workpiece like soda-lime glass The crack formation is less in hard wax as compared to the without hard wax. The material removal rate is less in hard wax than that without hard wax. The surface finishes with wax are more than that of the surface finish without wax.

(Continued)

TABLE 2.2 ***(Continued)***
Summary of Previous Rotary Ultrasonic Machining Research Initiatives

Author (Year)	Workpiece Material	Design of Exp. (DOE) Process Parameters (Input)	Machining Characteristics (Output)	Results/Conclusions
Tong et al. (2014)	Optical glass	Rotational speed (1,000–5,000 r/min) Feed rate (2–10 mm/min) Amplitude (2–10 μm) Ultrasonic frequency (27,900 Hz)	Surface roughness (SF)	The rotary machining process was reported as the surface roughness of the workpiece was reduced during the increase of spindle speed and feed rate. During machining, the surface roughness has a small amount of effect on amplitude.
Geng et al. (2014)	CFRP	Feed rate (0.33 mm/s) Spindle speed (5,000 r/min) Drilling depth (10 mm)	Tool life Drilling force	RUSM result, while machining on various input parameters, shows that compared to other parameters, drilling force is lower and tool life is enhanced by 28%.
Wang et al. (2018)	Glass ceramics	Spindle speed (3,000 rpm), feed rate (3 mm/min)	Residual stress Hole-size variation	Rotary ultrasonic results indicate that the machine holes increased gradually with the increase of feed rate. Moreover, the increase in speed of the spindle shows the residual stress tends to decrease.
Popli and Gupta (2018)	Advance ceramics (Al_2O_3)	Spindle rotation (3,000 rpm) Feed rate (0.010 mm/sec) Ultrasonic power rating (30%) Vibration frequency (20 kHz)	Chipping size	Rotary ultrasonic machining experimentally studied on Al_2O_3 ceramics material analyzes that input parameters have a significant effect on output results, indicating that the decrease of chipping size shows there is a reduction of spot length.
Singh et al. (2017)	UL-752 glass (polycarbonate bulletproof glass)	Concentration (20%–40%) Power rating (20 to 60%) Grit size (280–600) Tool material (high carbon steel) Frequency (20 kHz) Amplitude (25.3–25.8 μm)	Material removal rate (MRR) Tool wear rate (TWR) Surface roughness (SF)	Material removal rate and tool wear rate have a significant effect on the machining of glass material (UL-752). As a result, the increase in hardness of the material affects the increase of material removal rate. Moreover, the higher the grit size of the material, the higher the tool wear rate and material removal rate. Due to the fine grit size, it shows that there is an increase in surface roughness and a decrease in the removal of material as well as tool wear rate.

2.8 CONCLUSION

Rotary ultrasonic machining is typically referred to as a hybrid type of traditional ultrasonic and chemical-assisted ultrasonic machining process. Since there are so many different input machining techniques, the machining process is extremely difficult. Spindle speed, feed rate, vibration frequency, ultrasonic power, and grit size are important input machining process parameters that have a larger impact on the output machining parameters. Following a review of the output response literature, many conclusions include the following:

2.8.1 Material Removal Rate (MRR)

The different experimental investigation studies done on materials glass, ceramic, and composites on RUSM indicate that MRR increases with rotation speed, feed rate, and vibration frequency; however, grit size and ultrasonic power have little impact on MRR. The chemically assisted USM shows an increase of 200% MRR of macro drilling of glass.

2.8.2 Surface Roughness (SR)

According to a review of several studies on the machining of silicon carbides, carbon fiber composites, and alumina, surface roughness is extremely significantly linked with spindle speed and strongly related to feeding rate. Surface roughness is less influenced by vibration frequency and ultrasonic power. Moreover, ultrasonic power and vibration frequency have less impact on surface roughness.

2.8.3 Tool Wear Rate (TWR)

As per the study of the literature of tool wear during the machining of ceramic materials, CFRP, and silicon carbide, the most influential factor is feed rate. Increasing the feed rate causes more tool wear. The grain size also affects tool wear with high grain size resulting in more wear.

REFERENCES

Abdo, B.M.A., Darwish, S.M., Al-Ahmari, A.M., & El-Tamimi, A.M. (2013). Optimization of process parameters of rotary ultrasonic machining based on Taguchi Method. *Adv. Mater. Res.* 748, 273–280. https://doi.org/10.4028/www.scientific.net/AMR.748.273

Agarwal, S. (2015). On the mechanism and mechanics of material removal in ultrasonic machining. *Int. J. Mach. Tools Manuf.* 96, 1–14. https://doi.org/10.1016/j.ijmachtools.2015.05.006

Akhai, S., & Rana, M. (2022). Taguchi-based grey relational analysis of abrasive water jet machining of Al-6061. *Mater. Today Proc.* 65, 3165–3169. http://doi.org/10.1016/j.matpr.2022.05.361

Babbar, A., Jain, V., Gupta, D., & Agrawal, D. (2021b). Finite element simulation and integration of CEM43 °C and Arrhenius Models for ultrasonic-assisted skull bone grinding: A thermal dose model. *Med. Eng. Phys.* 90, 9–22.

Babbar, A., Jain, V., Gupta, D., Prakash, C., & Sharma, A. (2020b). Fabrication and machining methods of composites for aerospace applications. In *Characterization,*

Testing, Measurement, and Metrology (1st ed., pp. 109–124). Boca Raton, FL: CRC Press.

Babbar, A., Jain, V., Gupta, D., Prakash, C., Singh, S., & Sharma, A. (2020a). 3D bioprinting in pharmaceuticals, medicine, and tissue engineering applications. In *Advanced Manufacturing and Processing Technology* (1st ed., pp. 147–161). Boca Raton, FL: CRC Press.

Babbar, A., Jain, V., Gupta, D., Prakash, C., Singh, S., & Sharma, A. (2020c). Effect of process parameters on cutting forces and osteonecrosis for orthopedic bone drilling applications. In *Characterization, Testing, Measurement, and Metrology* (1st ed., pp. 93–108). Boca Raton, FL: CRC Press.

Babbar, A., Jain, V., Gupta, D., & Sharma, A. (2020d). Fabrication of microchannels using conventional and hybrid machining processes. In *Non-Conventional Hybrid Machining Processes* (1st ed., pp. 37–51). Boca Raton, FL: CRC Press.

Babbar, A., Rai, A., & Sharma, A. (2021a). Latest trend in building construction: Three-dimensional printing. *J. Phys. Conf. Ser.* 1950, 012007.

Babbar, A., Sharma, A., & Singh, P. (2022). Multi-objective optimization of magnetic abrasive finishing using grey relational analysis. *Mater. Today Proc.* 50, 570–575.

Baek, D.K., Ko, T.J., & Yang, S.H. (2013). Enhancement of surface quality in ultrasonic machining of glass using a sacrificing coating. *J. Mater. Proc. Technol.* 213(4), 553–559.

Bhosale, S.B., Pawade, R.S., & Brahmankar, P.K. (2014). Effect of process parameters on MRR, TWR and surface topography in ultrasonic machining of alumina-zirconia 83 ceramic composite. *Ceram. Int.* 40, 2831–12836. https://doi.org/10.1016/j.ceramint.2014.04.13712

Choi, J.P., Jeon, B.H., & Kim, B.H. (2007). Chemical-assisted ultrasonic machining of glass. *J. Mater. Proc. Technol.* 191(1–3), 153–156.

Churi, N. (2010). Rotary ultrasonic machining of hard-to-machine materials. Doctoral dissertation, Kansas State University.

Churi, N.J., Pei, Z.J., Shorter, D.C., & Treadwell, C. (2007). Rotary ultrasonic machining of silicon carbide: Designed experiments. *Int. J. Manuf. Technol. Manag.* 123, 284–298. https://doi.org/10.1504/IJMTM.2007.014154

Cong, W.L., Pei, Z.J., & Deines, T. (2010). Rotary ultrasonic machining of stainless steels: Empirical study of machining variables. *Int. J. Manuf. Res.* 5, 370. https://doi.org/10.1504/ijmr.2010.033472

Cong, W.L., Pei, Z.J., Feng, Q., Deines, T.W., & Treadwell, C. (2012). Rotary ultrasonic machining of CFRP: A comparison with twist drilling. *J. Reinf. Plast. Compos.* 31(5), 313–321.

Cong, W.L., Pei, Z.J., Mohanty, N., Van Vleet, E., & Treadwell, C. (2011). Vibration amplitude in rotary ultrasonic machining: A Novel measurement method and effects of process variables. *J. Manuf. Sci. Eng.* 133, 1–6.

Cong, W.L., Pei, Z.J., & Treadwell, C. (2014). Preliminary study on rotary ultrasonic machining of CFRP/Ti stacks. *Ultrasonic.* 54(6), 1594–1602.

Dhuria, G.K., Singh, R., & Batish, R.A. (2011). Ultrasonic machining of titanium and its alloys: A state of art review and future prospective. *Int. J. Mach. Mach. Mater.* 10(4), 326–355.

Geng, D., Zhang, D., Xu, Y., He, F., & Liu, F. (2014). Comparison of drill wear mechanism between rotary ultrasonic elliptical machining and conventional drilling of CFRP. *J. Reinf. Plast. Compos.* 33(9), 797–809.

Gong, H., Fang, F.Z., & Hu, X.T. (2010). Kinematic view of tool life in rotary ultrasonic side milling of hard and brittle materials. *Int. J. Mach. Tools Manuf.* 50, 303–307. https://doi.org/10.1016/j.ijmachtools.2009.12.006

Guzzo, P.L., Raslan, A.A., & DeMello, J.D.B. (2003). Ultrasonic abrasion of quartz crystals. *Wear.* 255, 67–77.

Hocheng, H., & Kuo, K.L. (2002). Fundamental study of ultrasonic polishing of mold steel. *Int. J. Mach. Tools Manuf.* 42(1), 7–13.

Jain, N.K., & Jain, V.K. (2001). Modeling of material removal in mechanical type advanced 82 machining processes: A state-of-art review. *Int. J. Mach. Tools. Manuf.* 41, 1573–1635. https://doi.org/10.1016/S0890-6955(01)00010

Jain, V.K. (2011). *Advanced Machining Processes.* Geneva: Inderscience Enterprises, p. 3.

Jiao, Y., Hu, P., Pei, Z.J., & Treadwell, C. (2005). Rotary ultrasonic machining of ceramics: Design of experiments. *Int. J. Manuf. Techno. Manag.* 7, 192–206. https://doi.org/10.1504/IJMTM.2005.006830

Kalia, G., Sharma, A., & Babbar, A. (2022). Use of three-dimensional printing techniques for developing biodegradable applications: A review investigation. *Mater. Today Proc.* 62, 346–352.

Kataria, R., Kumar, J., & Pabla, B.S. (2015). Experimental investigation into the hole quality in ultrasonic machining of WC-Co composite. *Mater. Manuf. Process.* 30, 921–933. https://doi.org/10.1080/10426914.2014.995052

Khanduja, P., Bhargave, H., Babbar, A., Pundir, P., & Sharma, A. (2021). Development of two-dimensional plotter using programmable logic controller and human machine interface. *J. Phys. Conf. Ser.* 1950, 012012.

Khoo, C.Y., Hamzah, E., & Sudin, I. (2008). A review on the rotary ultrasonic machining of advanced ceramics. *J. Mek.* 25, 9–23.

Kumar, S., Sudhakar, R.P., Goyal, D., & Sehgal, S. (2021). Process modelling for machining Inconel 825 using cryogenically treated carbide insert. *Met. Powder Rep.* 76, 66–74. http://doi.org/10.1016/j.mprp.2020.06.001

Kumar, V., & Singh, H. (2018). Regression analysis of surface roughness and micro-structural study in rotary ultrasonic drilling of BK7. *Ceram. Int.* 44(14), 16819–16827.

Kuo, K.L., & Tsao, C. (2012). Rotary ultrasonic-assisted milling of brittle materials. *T. Nonferr. Metal Soc.* 22, 793–800.

Legge, P. (1966). Machining without abrasive slurry. *Ultrasonic.* 4, 157–162.

Legge, P. (1964). Ultrasonic with a diamond drilling probe. *Ultrasonic.* 2, 1–4.

Liu, J., Zhang, D., Qin, L., & Yan, L. (2012). Feasibility study of the rotary ultrasonic elliptical machining of carbon fiber reinforced plastics (CFRP). *Int. J. Mach. Tools Manuf.* 53(2), 141–150.

Lv, D., Huang, Y., Wang, H., Tang, Y., & Wu, X. (2013). Improvement effects of vibration on cutting force in rotary ultrasonic machining of BK7 glass. *J. Mater. Proc. Technol.* 213(18), 1548–1557.

Nath, C., Lim, G.C., & Zheng, H.Y. (2012). Influence of the material removal mechanisms on hole integrity in ultrasonic machining of structural ceramics. *Ultrasonics.* 52(5), 605–613. https://doi.org/10.1016/j.ultras.2011.12.007

Ning, F.D., Cong, W.L., Pei, Z.J., & Treadwell, C. (2016). Rotary ultrasonic machining of CFRP: A comparison with grinding. *Ultrasonics.* 66, 125–132.

Pei, Z.J., & Ferreira, P.M. (1998). Modelling of ductile-mode material removal in rotary ultrasonic machining. *Int. J. Mach. Tools Manuf.* 38(23), 1399–1418.

Pei, Z.J., Ferreira, P.M., & Haselkorn, M. (1995). Plastic flow in rotary ultrasonic machining of ceramics. *J. Mater. Process Techno.* 48, 771–777.

Pei, Z.J., Khanna, N., & Ferreira, P.M. (1995). Rotary ultrasonic machining of structural ceramics—A review. *Ceram. Eng. Sci. Proc.* 16(1), 259–278.

Pei, Z.J., Prabhakar, D., Ferreira, P.M., & Haselkorn, M. (1994). Rotary ultrasonic drilling and milling of ceramics. *Urbana.* 51, 61801.

Pérez, P., Royer, R., Merson, E., Lockwood, A., Ayvar-Soberanis, S., & Marshall, M.B. (2016). Influence of workpiece constituents and cutting speed on the cutting forces developed in the conventional drilling of CFRP composites. *Compos. Struct.* 140, 621–629. https://doi.org/10.1016/j.compstruct.2016.01.008

Popli, D., & Gupta, M. (2018). A chipping reduction approach in rotary ultrasonic machining of advance ceramic. *Mater. Today Proceed.* 5(2), 6329–6338.

Prakash, C., Kumar, V., Mistri, A., Sharma, A., Uppal, A.S., Babbar, A., & Pathri, B.P. (2021). Investigation of functionally graded adherents on failure of socket joint of FRP composite tubes. *Materials.* 14, 6365.

Rampal, R., Goyal, T., Goyal, D., Mittal, M., Dang, R.K., & Bahl, S. (2021). Magneto-rheological abrasive finishing (MAF) of soft material using abrasives. *Mater. Today Proc.* 45, 51140–5121. http://doi.org/10.1016/j.matpr.2021.01.629

Rana, M., & Akhai, S. (2022). Multi-objective optimization of Abrasive water jet Machining parameters for Inconel 625 alloy using TGRA. *Mater. Today Proc.* 65, 3205–3210. http://doi.org/10.1016/j.matpr.2022.05.374

Sarwade, A. (2010). Study of micro rotary ultrasonic machining. M.S thesis, The Graduate College at the University of Nebraska. https://digitalcommons.unl.edu/imsediss/6/

Sharma, A., Babbar, A., Jain, V., & Gupta, D. (2018b). Enhancement of surface roughness for brittle material during rotary ultrasonic machining. *MATEC Web of Conf.* 249, 01006.

Sharma, A., Babbar, A., Tian, Y., Pathri, B.P., Gupta, M., Singh, R. (2022c). Machining of Ceramic Materials: A state of the art review. *Int. J. Interact. Des. Manuf. (IJIDeM)*, 1–21. https://doi.org/10.1007/s12008-022-01016-7

Sharma, A., Fidan, I., Huseynov, O., Ali, M.A., Alkunte, S., Rajeshirke, M., Gupta, A., Hasanov, S., Tantawi, K., Yasa, E., Yilmaz, O., Loy, J., & Popov, V. (2023b). Recent inventions in additive manufacturing: Holistic review. *Inventions.* 8(4), 103. https://doi.org/10.3390/inventions8040103

Sharma, A., & Jain, V. (2020). Experimental investigation of cutting temperature during drilling of float glass specimen. In *IOP Conference Series: Materials Science and Engineering* (Vol. 715, No. 1, p. 012050). IOP Publishing. https://doi.org/10.1088/1757-899X/715/1/012050

Sharma, A., Jain, V., & Gupta, D. (2018a). Characterization of chipping and tool wear during drilling of float Glass using rotary ultrasonic machining. *Measurement.* 128, 254–263.

Sharma, A., Jain, V., & Gupta, D. (2019a). Comparative analysis of chipping mechanics of float glass during rotary ultrasonic drilling and conventional drilling: For multi-shaped tools. *Mach. Sci. Technol.* 23, 547–568.

Sharma, A., Jain, V., & Gupta, D. (2019b). Multi-shaped tool wear study during rotary ultrasonic drilling and Conventional drilling for amorphous solid. *Proc. Inst. Mech. Eng. Part E. J. Process Mech. Eng.* 233, 551–560.

Sharma, A., Jain, V., & Gupta, D. (2019c). Tool wear analysis while creating blind holes on float glass using conventional drilling: A multi-shaped tools study. In *Advances in Manufacturing Processes: Select Proceedings of ICEMMM 2018* (pp. 175–183). Singapore: Springer. https://doi.org/10.1007/978-981-13-1724-8_17

Sharma, A., Jain, V., & Gupta, D. (2021a). Effect of pre and post tempering on hole quality of float glass specimen: For rotary ultrasonic and conventional drilling. *Silicon.* 13, 2029–2039.

Sharma, A., Jain, V., & Gupta, D. (2022a). Mathematical approach on chipping volume estimation generated during rotary ultrasonic drilling for float glass. *Proc. Natl. Acad. Sci. India Sect. A Phys. Sci.* 92, 285–291.

Sharma, A., Jain, V., Gupta, D., & Babbar, A. (2020). A review study on miniaturization. In *Advanced Manufacturing and Processing Technology* (1st ed., pp. 111–131). Boca Raton, FL: CRC Press.

Sharma, A., Jain, V., Gupta, D., & Babbar, A. (2020d). A review study on miniaturization. In *Advanced Manufacturing and Processing Technology* (1st ed., pp. 111–131). Boca Raton, FL: CRC Press.

Sharma, A., Kalsia, M., Uppal, A.S., Babbar, A., & Dhawan, V. (2022b). Machining of hard and brittle materials: A comprehensive review. *Mater. Today Proc.* 50, 1048–1052.

Sharma, A., Kumar, V., Babbar, A., Dhawan, V., & Kotecha, K. (2021b). Experimental investigation and optimization of electric discharge machining process parameters using grey-fuzzy-based hybrid techniques. *Materials* 14(19), 5820.

Sharma, A., Parikh, P.A., Roy, D., Joshi, K., & Trivedi, R. (2023c). Performance evaluation of an indigenously-designed high-performance dynamic feeding robotic structure using advanced additive manufacturing technology, machine learning and robot kinematics. *Int. J. Interact. Des. Manuf. (IJIDeM)*, 1–29.

Sharma, A., Sandhu, H.S., Goyal, D., Goyal, T., Jarial, S., & Sharda, A. (2023a). Sustainable development in cold gas dynamic spray coating process for biomedical applications: Challenges and future perspective review. *Int. J. Interact. Des. Manuf. (IJIDeM)*, 1–17. https://doi.org/10.1007/s12008-023-01474-7

Sharma, V.K., Rana, M., Singh, T., Singh, A.K., & Chattopadhyay, K. (2021c). Multi-response optimization of process parameters using Desirability Function Analysis during machining of EN31 steel under different machining environments. *Mater. Today Proc.* 44, 3121–3126. http://doi.org/10.1016/j.matpr.2021.02.809

Sindhu, D., Thakur, L., & Chandna, P. (2018). Multi objective optimization of rotary ultrasonic machining parameters for quartz glass using Taguchi-Grey relational analysis (GRA). *Silicon*, 1–12.

Singh, B.P., Singh, J., Bhayana, M., & Goyal, D. (2021). Experimental investigation of machining nimonic-80A alloy on wire EDM using response surface methodology. *Met. Powder Rep.* 76, 9–17. http://doi.org/10.1016/j.mprp.2020.12.001

Singh, B.P., Singh, J., Bhayana, M., Singh, K., & Singh, R. (2022). Experimental examination of the machining characteristics of Nimonic 80-A alloy on wire EDM. *Mater. Today Proc.* 69, 291–296. http://doi.org/10.1016/j.matpr.2022.08.537

Singh, J., Singh, C., & Singh, K. (2023). Rotary ultrasonic machining of advance materials: A review. *Mater. Today Proc.* http://doi.org/10.1016/j.matpr.2023.01.159

Singh, K.J., Ahuja, I.S., & Kapoor, J. (2017). Grey relational analysis of chemical assisted USM of polycarbonate bullet proof (UL-752) & acrylic heat resistant (BS-476) glass. *Am. J. Mech. Eng.* 5(3), 94–109.

Singh, R.P., & Singhal, S. (2016). Rotary ultrasonic machining: A review. *Mater. Manuf. Process.* 6914, 0–119. https://doi.org/10.1080/10426914.2016.1140188

Singh, R.P., & Singhal, S. (2018). Experimental investigation of machining characteristics in rotary ultrasonic machining of quartz ceramic. *Proc. Inst. Mech. Eng. Part L J. Mater. Des. Appl.* 232(10), 870–889.

Tabatabaei, S.M.K., Behbahani, S., & Mirian, S.M. (2013). Analysis of ultrasonic assisted machining (UAM) on regenerative chatter in turning. *J. Mater. Proc. Technol.* 213(19), 418–425.

Tong, Y., Chen, J., Sun, L., Yu, X., & Wang, X. (2014, August). Research on the processing technology of elongated holes based on rotary ultrasonic drilling. In *7th International Symposium on Advanced Optical Manufacturing and Testing Technologies: Advanced Optical Manufacturing Technologies* (Vol. 9281, pp. 346–353). USA:SPIE.

Vinod, K., & Khamba, J.S. (2010). An investigation into the ultrasonic machining of co-based super alloy using the Taguchi approach. *Int. J. Mach. Mach. Mater.* 7, 230–243. https://doi.org/10.1504/IJMMM.2010.033068

Wang, H., Pei, Z.J., & Cong, W.L. (2020a). A feeding-directional cutting force model for end surface grinding of CFRP composites using rotary ultrasonic machining with elliptical ultrasonic vibration. *Int. J. Mach. Tools Manuf.* https://doi.org/10.1016/j.ijmachtools.2020.10354

Wang, H., Zhang, D., Li, Y., & Cong, W. (2020b). The effects of elliptical ultrasonic vibration in surface machining of CFRP composites using rotary ultrasonic machining. *Int. J. Adv. Manuf. Technol.* 106, 5527–5538. https://doi.org/10.1007/s00170-020-04976-w

Wang, J., Feng, P., & Zhang, J. (2018). Reducing edge chipping defect in rotary ultrasonic machining of optical glass by compound step-taper tool. *J. Manuf. Process.* 32, 213–221.

Wang, J., Zha, H., Feng, P., & Zhang, J. (2016). On the mechanism of edge chipping reduction in rotary ultrasonic drilling: A novel experimental method. *Precis Eng.* 44, 231–235.

Wang, Q., Cong, W., Pei, Z.J., Gao, H., & Kang, R. (2009). Rotary ultrasonic machining of dihydrogen phosphate (KDP) crystal: An experimental investigation on surface roughness. *J. Manuf. Proc.* 11, 66–73.

Ya, G., Qin, H.W., Yang, S.C., & Xu, Y.W. (2002). Analysis of the rotary ultrasonic machining mechanism. *J. Mater. Process Techno.* 129, 182–185. https://doi.org/10.1016/S0924-0136(02)00638-6

Zhang, C., Cong, W., Feng, P., & Pei, Z. (2014). Rotary ultrasonic machining of optical K9 glass using compressed air as coolant: A feasibility study. *Proc. Inst. Mech. Eng. Part. B. J. Eng. Manuf.* 228, 504–514. https://doi.or g/10.1177/0954405413506195

3 Effect of Minimum Quantity Lubrication (MQL) Method on Machining Characteristics for Ductile Substrates

A Future Direction

Ankit Sharma, Atul Babbar, Kamaljeet Singh, Anoop Kumar Singh, Naveen Mani Tripathi, Dhaval Jaydev Kumar Desai

3.1 OUTLOOK TO THE MACHINING PROCESS

In many manufacturing procedures, machining is the most crucial step. At some point throughout their manufacture, the majority of the items need to be machined. A metal-removal procedure called machining is used to give the workpiece the finishing touch and needed dimensional accuracy. This primary procedure is for changing the geometry of a workpiece. The methods for removing material are divided into three categories: cutting, grinding, and other specialized methods. Cutting operations remove material by creating ribbonlike structures known as chips with 0.025 to 2.5 mm of thickness, while grinding operations typically produce chips with a size ranging from 0.0025 mm to 0.25 mm as a result of breaking down the removed material into smaller particles (Bruni et al., 2006). In the manufacturing sector, a variety of machining techniques are used to remove material over the substrate periphery in order to produce a finished good (Sahoo et al., 2008). Turning, milling, and drilling are three of the many machining techniques that are used to remove metal.

In any process that comprises metal machining, the cutting fluid plays an essential part. It does this by eliminating chips from the cutting zone, cooling the surface of the material and the cutting tool, and lubricating the interaction between the tool and the workpiece. Instead, if the cutting fluid is not used properly or disposed of incorrectly, it can have a negative impact on both human health and the environment. Researchers

DOI: 10.1201/9781003327905-3

have been concentrating their efforts on a technique known as minimum quantity lubrication (MQL) because it reduces the amount of coolant that is used by spraying a combination of compressed air and cutting fluid in an optimized means rather than using flood cooling. This method is one of many that are available for the application of the coolant. Because it satisfies all the criteria for what constitutes "green" machining, the MQL approach has been found to be an effective solution. On the other hand, MQL is making its way toward the advanced method of machining processes. This technique is meant to reduce the amount of cutting fluids and maintain the optimum cutting force that provides the desired surface quality. The major outcome of the study stated that the concentration of nanoparticles in the base fluid has a positive effect on the tribological properties of the fluid. Since the tribological properties of the cutting fluid enhanced with nanoparticles are superior to those of the base fluid, some investigations stated that the use of MQL and MQL that had been combined with water proved to be effective in enhancing the surface finish of the object and reducing the tool wear rates. Because of its low cost, high level of protection for the environment, and long-term viability in the environment, MQL has become increasingly popular in the field of machining in recent years. It has been shown in some tests to be capable of providing good cooling and lubrication, which presents an opportunity for ceramic cutting tools to test the process (Singh et al., 2020b).

3.2 MACHINING CONDITIONS

Three fundamental categories of machining environments are deployed in the industry, and they are as follows:

3.2.1 Flood Machining

The cutting fluid (oil and water) is utilized in this type of machining condition in a large amount (between 2,000 and 4,000 litres per hour) during the interaction between the substrate specimen and tool to extract the chips and simultaneously cool the substrate specimen and tooling interaction.

3.2.2 Dry Machining

For the sake of conserving cutting fluids during flood machining, dry machining is deployed instead since it eliminates the need for cutting fluid, a major source of pollution, and other harmful by-products of the machining process. Surface quality and material clearance rate won't improve with this method either. In order to improve output responsiveness and the surrounding environment while falling the quantity of coolant and lubricant required, the MQL method is used.

3.2.3 MQL Machining

Numerous issues with health and the environment are brought about by the overuse of machining fluid at the time of flood cooling. The minimal quantity lubrication (MQL) method is utilized to cool the tool-workpiece contact while simultaneously reducing coolant consumption. A spray nozzle is needed for MQL mist cooling in order to combine high-pressure air with a modest volume of cutting fluid (between 50 and 500 ml/hr). The oil coolant is broken down into incredibly small particles by this spray nozzle, which flows inside the high-pressure air jet. The tool-work interaction is the target of this mist lubrication (Sales et al., 2001).

3.2.4 Minimum Quantity Lubrication (MQL) Technique

Cutting fluid overuse has an undesirable influence on the atmosphere and social health during both its use and removal (Weinert et al., 2004). Therefore, it is best to avoid using cutting fluids excessively. Other cooling approaches are being developed in addition to standard cooling methods to take the heat away from the tool-work contact (Sreejith & Ngoi, 2000). The temperature in the cutting zone can be controlled in a variety of ways, including by using cooling techniques including flood cooling, minimum quantity lubrication (MQL), solid lubricants, high-pressure coolants, compressed air/gas coolants, cryogenic cooling, and so on (Sharma et al., 2009). Numerous experts have looked into the impact of different cooling methods combined with lubricant oil. They have come to the conclusion that the minimal quantity lubrication (MQL) technology excels over alternative cooling systems due to its low cutting fluid consumption and environmentally friendly machining (Lawal et al., 2013).

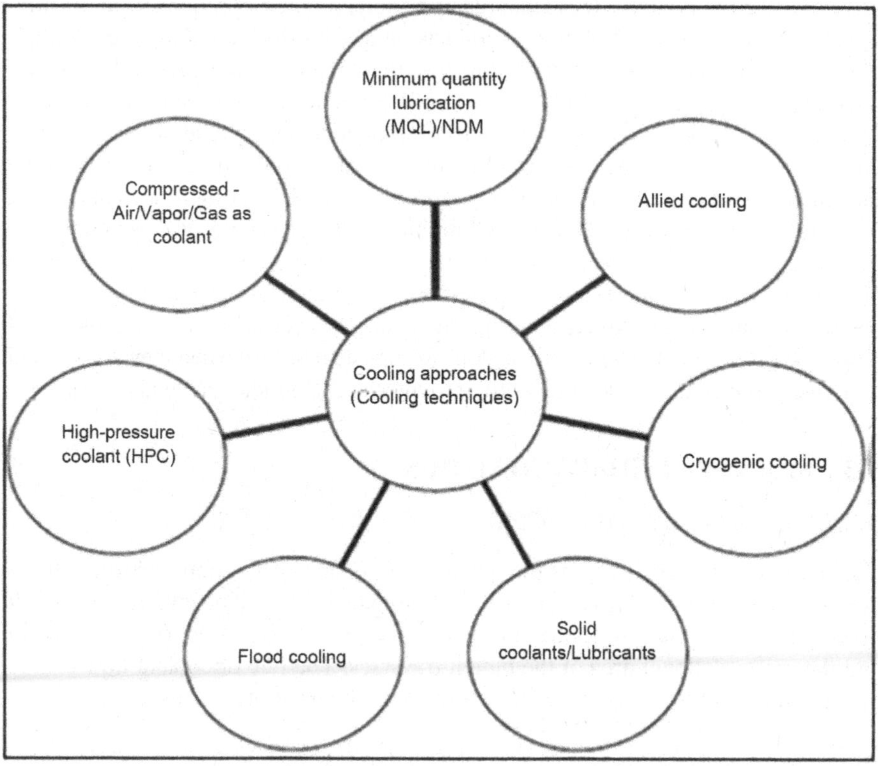

FIGURE 3.1 Different cooling techniques.

Source: Sharma et al. (2009)

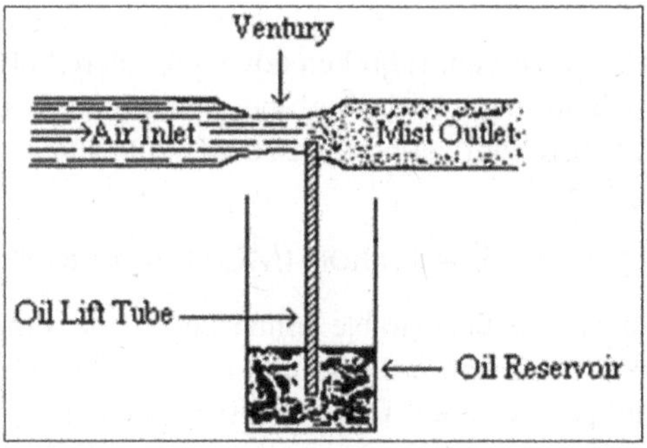

FIGURE 3.2 Principle of MQL oil mist generation.

Source: Varadarajan et al. (2002)

Due to its benefits in protecting the atmosphere and human well-being by deploying less cutting lubricants, the MQL approach is typically used on turning, milling, and drilling operations in the manufacturing area (Bhowmick & Alpas, 2011; Attanasio et al., 2006; Rahman et al., 2002). The MQL methodology encompasses the compensations of both the dry technique and flood cooling technique due to the mixing of lubricant and air. MQL has several advantages for the industrial sector, including lowering manufacturing costs, being environmentally friendly, addressing issues relating to human health, and addressing regulatory requirements (Braga et al., 2002).

The MQL method involves mixing a little amount of cutting fluid or lubricant with air to produce droplets. As shown in Figure 3.2, a nozzle (the interface between the tool and the workpiece) is used to spray these drops into the cutting zone at the proper pressure. The reservoir for the cutting fluid, the discharge nozzle, etc., are all part of this system. The MQL system takes advantage of the venturi effect to remove cutting fluid from the lubricating reservoir. The cutting fluid is atomized into microdroplets by passing pressurized air through a mixing chamber. These droplets act as a coolant and lubricant and penetrate deeply into the tool-workpiece interface when they are sprayed as a mist in the cutting zone (tool-workpiece interface) (Varadarajan et al., 2002).

3.3 MQL INPUT CHARACTERISTICS

3.3.1 Coolant Flow Rate (CFR)

Coolant flow rate (CFR) means the quantity of coolant that is pumped through the nozzle and into the cutting region in a given amount of time. The amount of coolant flowing from the reservoir to the cutting region is reliant on the lubricant's velocity and the cross-sectional area of the pipe it traverses. The rate of coolant flow can be quantified in millilitres per second, millilitres per hour, or litres per second.

3.3.2 Nozzle Tool Distance

The nozzle tool distance refers to the physical separation among the nozzle and the milling insert (milling tool). Surface roughness rises with an intensification in nozzle

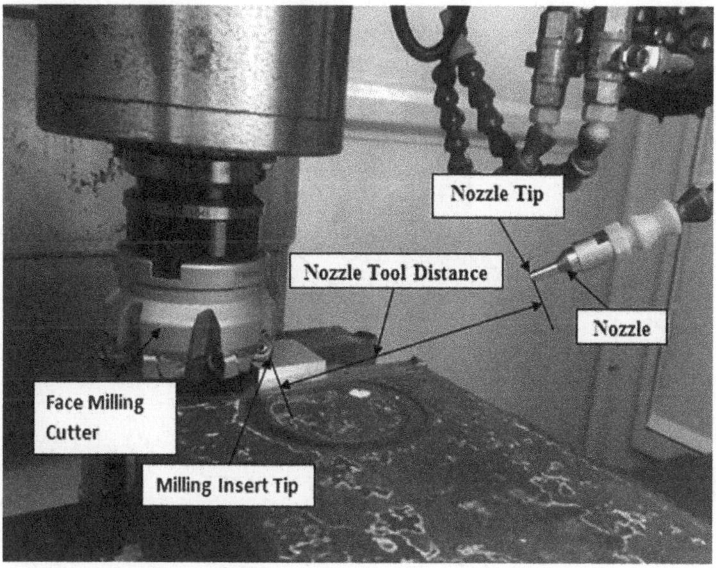

FIGURE 3.3 Real-time illustration of nozzle tool distance (NTD) on VMC.

Source: Singh et al. (2020a)

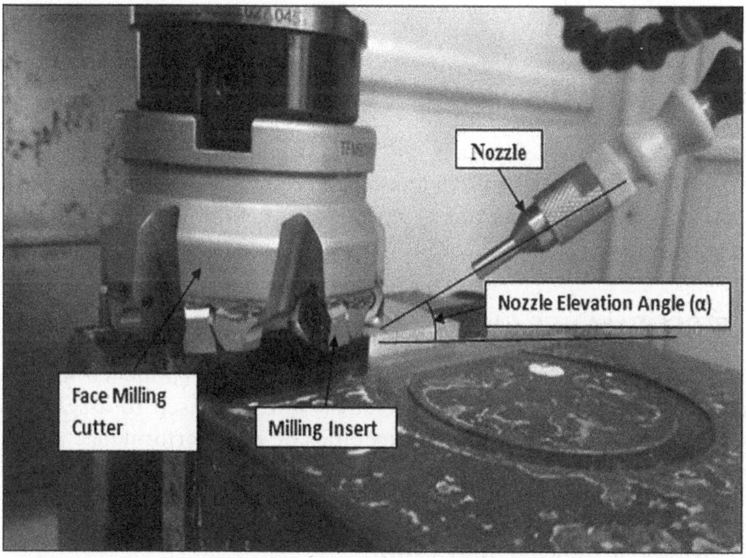

FIGURE 3.4 Real-time illustration of nozzle elevation angle (NEA) on VMC.

Source: Singh et al. (2020a)

tool distance up to a specific point and vice versa. Additionally, the form and diameter of the cutter are influenced by the nozzle tool distance. Both a graphic and a definite illustration of the nozzle tool distance (NTD) on a VMC are presented in Figures 3.3 and 3.5, respectively.

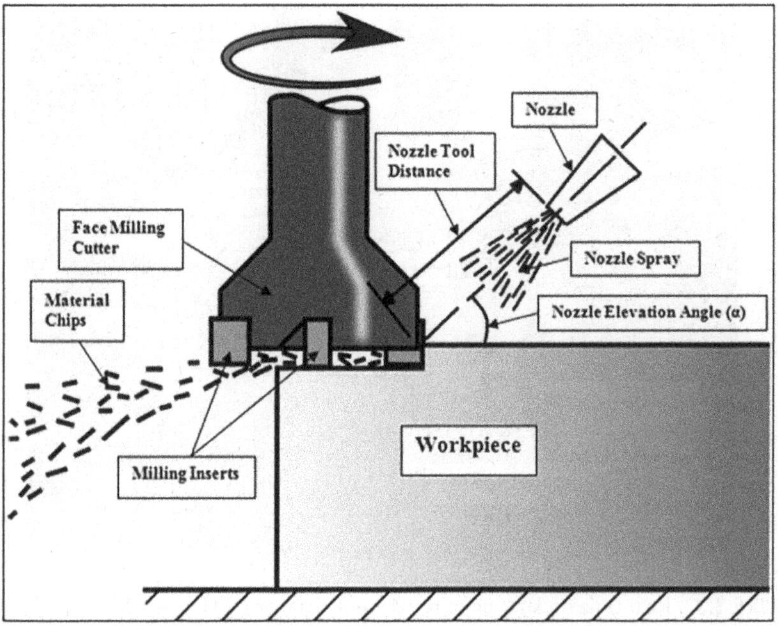

FIGURE 3.5 Machining mechanism using MQL based milling operation.

Source: Singh et al. (2020a)

3.3.3 Nozzle Elevation Angle (NEA)

The angle among the nozzle tip and milling insert tip is known as the "nozzle elevation angle" (NEA) (milling tool). Degree (°) units are used to express nozzle elevation angle (NEA). Both a visual and a real representation of the nozzle elevation angle are illustrated in Figures 3.4 and 3.5, respectively.

3.4 PERFORMANCE CHARACTERISTICS

Performance characteristics, sometimes referred to as eminence features such as surface roughness and material removal rate, are used to assess how well the milling process is working. The specifics of each performance attribute are covered here.

3.4.1 Material Removal Rate (MRR)

The material removal rate is definite as the rate at which material is detached from the workpiece (MRR). According to Equations 3.1 and 3.2, the material removal rate can be considered using the machining time and the difference in gram weight between the starting and finishing weights of the specimens (i.e. before and after machining). It can also be calculated using the difference in volume between the starting and finishing volumes. Material removal rate, which indicates how quickly or slowly the machining rate is, is a crucial performance measure in production rate. Therefore, achieving the needed production rate with the appropriate surface smoothness and precision is the industry's key goal. Many different aspects of the machining process (depth of cut, cutting velocity, etc.) can affect the material removal rate (MRR). Parameters in MQL code, beyond those used for machining inputs, include crucial variables that establish the rate of material removal. In any operation, the material removal rate (MRR) is the quantity of material removed from the workpiece in a given time period. In theory, it can be determined with Eq. 3.1. Weighing the workpiece before and after machining is essential, especially when the machining time is already established. MRR can also be premeditated by means of Eq. 3.2.

$$MRR = w \times d \times f \tag{3.1}$$

Where w = width of cut, d = depth of cut, and f = feed rate, MRR = material removal rate

$$MRR = \frac{W_i - W_f}{t} \tag{3.2}$$

Where W_i = initial weight of specimen before machining, W_f = final weight of specimen after machining, t = machining time

3.4.2 Surface Roughness (SR)

In the modern world, performance qualities of the products, such as surface roughness and dimensional accuracy, are given more attention. Even though the product's dimensions are well within the tolerance zone, there are many chances that the surface finish will cause the finished product to be rejected. A totally flat and smooth surface cannot be produced throughout any manufacturing process. Surface roughness or surface finish refers to the many small abnormalities as a result of the unavoidable conditions (noise factors) during machining. But low surface roughness components are always needed, especially when there is relative motion between two moving pieces (Behera et al., 2014).

The life of the part and the amount of electricity required to operate it are both impacted by surface roughness value. High surface roughness causes the mechanism's temperature to increase, which further increases the need for lubricating oil in the manufacturing industries (Sharma et al., 2018a). Surface roughness is influenced by the following factors: material, cutting tool condition, cooling method, vibration, machining settings, and fixture.

3.4.3 Surface Texture

Surface roughness is the degree to which the actual surface deviates from the idealized one. Surface roughness, waviness, and form are the three basic categories into which deviations fall. These surfaces' surface roughness profiles and these surfaces are created by superimposing different order deviations. The average roughness of a surface is calculated by averaging the absolute roughness departure values from the surface mean line over the assessment length. These averages may be above or below the geometric center of the plane. In many cases, the average roughness is determined by averaging the roughness measurements taken along the entire length under consideration. This is the typical parameter that has been used for a long time to measure surface roughness. It is sized in millimeters (Singh et al., 2014; Al-Attar et al., 2013). Figure 3.6 (a) and (b) illustrates the different terms related to surface texture and surface profile.

3.4.3.1 Roughness

Small deviations (minor imperfections) from the primary machined surface are referred to as roughness. Surface roughness is the word used to describe surface abnormalities of finer spacing.

3.4.3.2 Waviness

Waviness is a measure of how smooth the surface is. Structure has a role in its occurrence. Waviness is a type of surface irregularity that has a larger spacing or deviates from the primary machined surface.

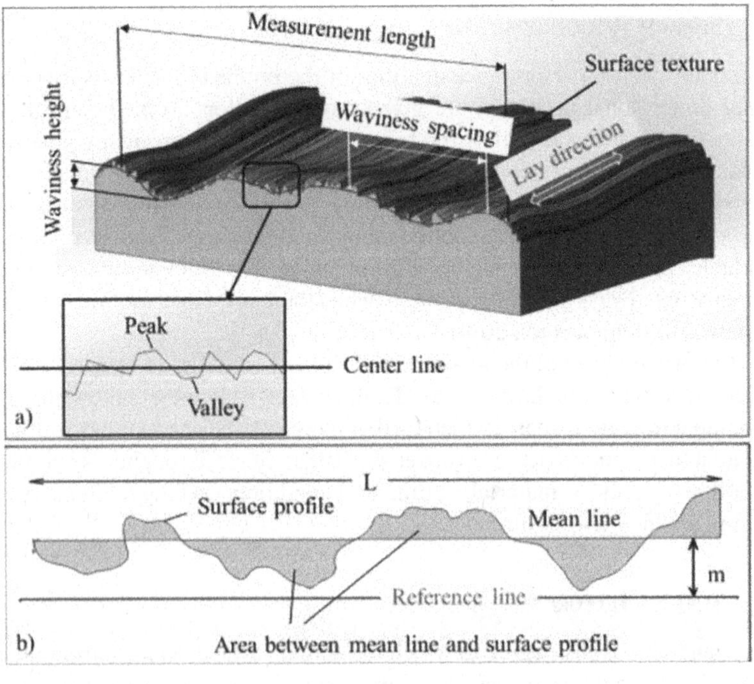

FIGURE 3.6 A) Terminologies of surface texture and b) surface profile.

Source: Kalami and Urbanic (2021)

3.4.3.3 Flaws

Cracks and scratches are examples of flaws that can appear on a product's surface owing to imperfections in the manufacturing process.

3.4.3.4 Lays

Lays depicts the surface patterns of a component as a consequence of the direction of the machining process. The machining process for a given component determines whether the lay pattern will be horizontal, vertical, or round (Singh et al., 2014; Eman et al., 2013).

3.5 REVIEW ON THE PERFORMANCE CHARACTERISTICS PARAMETERS

3.5.1 Cost of Cutting Fluid

Improvements in tool life and surface finish have resulted from the usage of cutting fluid to cool the machining process since the beginning of the twentieth century. Several kinds of cutting fluids have also been used for this and similar objectives. Several efforts have been made in the last ten years to diminish the quantity of cutting fluids used in manufacturing because of the costs of the fluids, environmental concerns, concerns about human health, and other factors. In 1992, the amount of waste soluble oil produced by German businesses accounted for around 1,151.312 tonnes, or about 60% of all the lubricants utilized in manufacturing. This sum of money, which ranges from 7.5% to 17% of manufacturing costs per part, is considerable and is even greater than the costs associated with tooling, according to Heisel et al. (1998), Klocke and Eisenblätter (1997), and Kalhofer (1997).

Cutting fluids are used globally, where 640 million gallons are used worldwide and 100 million gallons are deployed in the USA, according to research by Marksberry and Jawahir (2008). The basic functions of cutting fluids are to eliminate chips from the cutting area, to make the workpiece's surface more even and flat, to reduce friction among the tool and the workpiece, and to reduce the amount of heat generated by the cutting process. Even while these consequences are positive, the negative effects outweigh the positive ones, causing us to reconsider how we achieve these goals.

According to Khan et al. (2009), the cost of applying and arranging of cutting fluids in the production of automotive parts was around 17% of the whole cost. Giving to research by Sharma et al. (2016), the cost of cutting fluids accounts for roughly 16% to 20% of the manufacturing sector's overall cost of production. Figure 3.7 and Figure 3.8 depict the studies which are focused on manufacturing costs, machining costs, and the structure of coolant cost (Brinksmeier et al., 1999; Nasir, 1998; Sanchez et al., 2010).

3.5.2 Environment and Human Health

According to Chakraborty et al. (2008), many international organizations have all demanded that cutting fluid usage be decreased and that manufacturing industries everywhere be given a safe working environment.

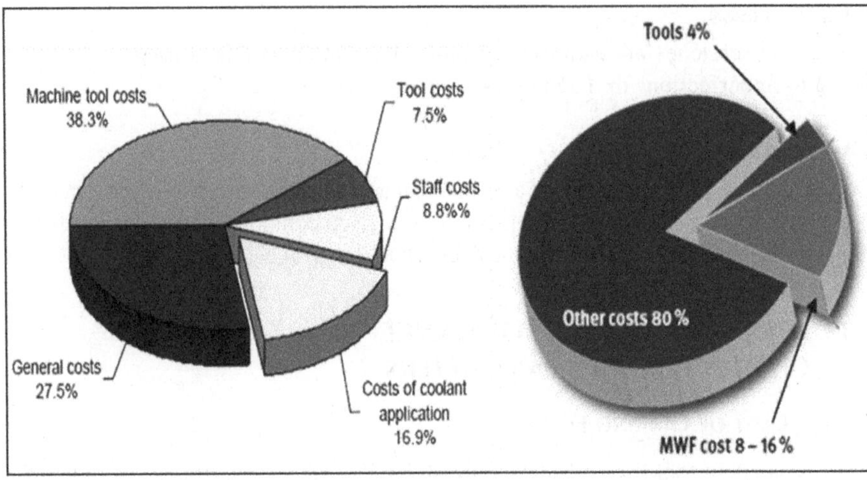

FIGURE 3.7 Demonstration of manufacturing costs, machining costs, and the assembly of coolant cost.

Source: Brinksmeier et al. (1999); Nasir (1998); Sanchez et al. (2010)

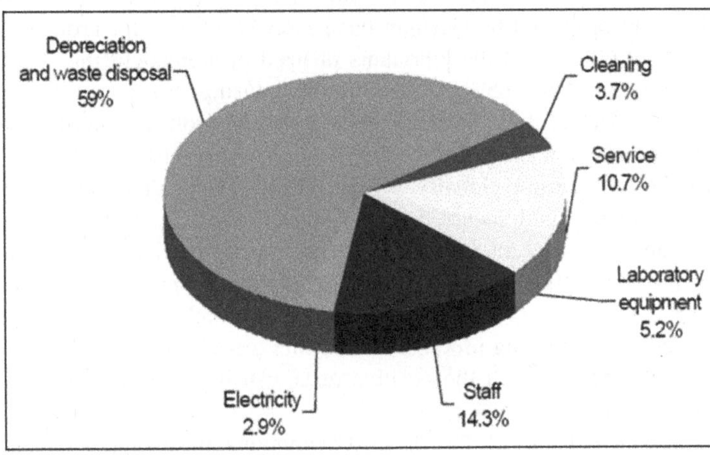

FIGURE 3.8 The chart shows the structure of coolant costs in different categories.

Source: Brinksmeier et al. (1999); Nasir (1998)

Howes et al. (1991) and Byrne and Scholta (1993) found that the use and removal of cutting fluids resulted in a number of environmental issues, including soil contamination, water pollution, air pollution, and environmental pollution. Greaves et al. (1997) looked into the relationship between cutting fluid aerosol exposure and symptoms of asthma, chronic conditions, and chest infections (MWFs). According to Sales et al. (2001) and the National Institute of Occupational Safety and Health (NIOSH), 1.2 million workers worldwide have gone through the harmful properties of lubricants during application and disposal, which over time resulted in breathing issues.

So as to make the metal-cutting process additional financially and ecologically sound, the various sustainable strategies in machining were acknowledged by Behera et al. (2014), who conducted an investigation into a study of all sustainable manufacturing approaches in machining for environmentally friendly manufacturing.

The minimal quantity lubrication strategy has gained popularity in the metal-cutting industries due to its low environmental impact. When dry or flood machining can't be justified for economic or ecological reasons, MQL technology is employed to prevent waste and pollution while saving money. Cutting fluids are sprayed directly into the point of contact among the tool and the workpiece.

3.5.3 Workpiece Specification, Mode of Cooling, and Other Output Characteristics

After conducting trials on a 7% Si aluminium alloy (SAE 323), Braga et al. (2002) determined that the MQL approach produced holes of equal or superior quality to flood cooling. Cutting speed (300 m/min), feed (0.1, 0.2 mm/rev), and MQL flow rate (10 ml/hr) are only some machining levels and ranges that can be selected.

Sharma and Sidhu (2014) used vegetable oil as a lubricant to study the differences between dry and MQL cooling methods for AISI D2 steel. Together, MQL/NDM offers a greener means of production and helps improve several aspects of product quality (such as surface polish).

Tasdelen et al. (2008) looked at the surface finishes of 272 and 315 drilled holes. Overall, the Ra and Rz values from the 15 ml/h MQL technique are superior to those from the flood drilling method.

As illustrated in Figure 3.9, Khan et al. (2009) investigation into the impact of turning operations on AISI 9310 alloy steel utilizing cutting fluid based on vegetable

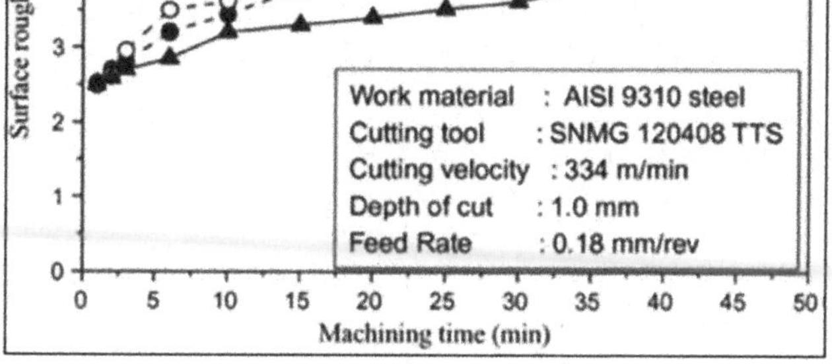

FIGURE 3.9 Surface roughness under different environments (dry, wet, and MQL) turning.

Source: Khan et al. (2009)

oil discovered that the MQL approach generated the best surface finish. MQL flow rate and other machining factors are some of the input parameters, along with their respective ranges.

Figure 3.10 displays the results of a comparison between the surface roughness generated by wet, dry, and MQL lubrication methods across a range of cutting speeds and cutting times. Surface roughness (Ra) inclines to reduction with growing cutting time when utilizing flood or wet cooling systems; however, a minor shift may be seen when employing dry cutting. The Ra values for surfaces finished using the MQL method were found to be comparable to, or even lower than, those produced by wet and dry cutting.

When turning operations were compared to wet or flood turning, Hwang and Lee (2010) used the MQL technique. It was determined that MQL turning performs better than wet turning in terms of surface roughness. Input machining parameters for turning with MQL that work best together are feed rate of 0.01 mm/rev, cutting speed of 361 m/min, depth of cut of 0.1 mm, and nozzle diameter of 6 mm.

According to Tosun and Huseyinoglu (2010), TiCN cutting tool and flooded cooling gave equal results when used to mill aluminium alloy (Al-7075), but MQL milling produced a better surface polish using different input parameters and their respective ranges.

According to Li and Chou (2010), using the MQL approach with the same cutting speed and feed rate as a flooded cooling system resulted in decreased surface roughness at spindle rotational speed ranges from 2,000 to 4,000 rpm and feed rate of 1 to 2 mm/rev. The MQL flow rate was reported as 1.88 and 7.5 ml/hr. These are among the different input parameters and their respective ranges.

Because MQL milling produces results that are similar to those of wet milling in terms of surface roughness and cutting power, Fratila and Caizar (2011) have looked into whether it may be used in place of wet milling at a speed of 150.72 m/min with 0.1 mm/rev of feed rate. The MQL flow rate was considered as 30 ml/hr and the depth of cut as 1 mm, which has some of the input parameters.

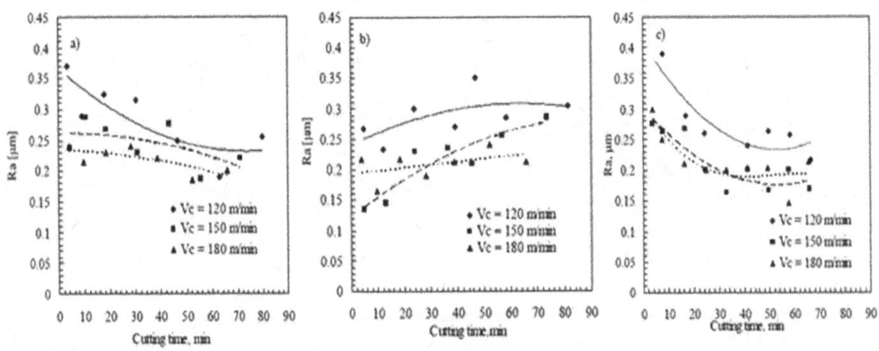

FIGURE 3.10 Surface roughness versus cutting time under wet (a), dry (b), and MQL (c) at Vc =120, 150, and 180 m/min.

Source: Bruni et al. (2008)

Hadad and Sadeghi (2013) found that MQL turning, coupled with wet and dry turning, may achieve the same depth of cut when working with AISI 4140 alloy steel. MQL was shown to have the best surface finish compared to both dry and wet machining with a variety of nozzle placements and orientations.

Turning with soluble oil cutting fluid was deliberate by Amrita et al. (2014), who found that in comparison to flooded cooling systems, cutting temperatures were 25% higher, cutting forces were 54% higher, surface roughness was 30% higher, and tool wear was 71% higher when using the MQL cooling system.

Ti-48Al-2Cr-2Nb, an intermetallic alloy, was tested to see how milling and turning affected it (Priarone et al., 2014). Although employing wet, dry, or MQL cooling methods. After conducting tests, it was shown that MQL cooling yields the smoothest surface (0.58 m), beating out both wet cooling conditions (0.74 m) and dry cooling conditions (0.98 m, 0.82 m). Surface roughness as a meaning of cooling condition is publicized in Figure 3.11.

The MQL factors in turning were optimized by means of the Taguchi and Grey relational technique by Sarıkaya and Güllü (2015). Vegetable oil cutting fluid is used in experiments to decrease tool wear and surface roughness. For maximum efficiency, the fluid flow rate is 180 ml/hr and the cutting speed is 30 m/min.

According to research by Uysal et al. (2015), the MQL process produced superior surface roughness outcomes than dry milling. At a flow rate of 40 ml/hr, the least surface roughness that could be achieved through milling was leisurely to be 0.865 m. By comparing the MQL method to dry milling, the improvements in surface finish values were 8.8% for 20 ml/hr and 22.5% for 40 ml/hr.

Using the MQL lubrication approach, Hassanpour et al. (2016) experimented on AISI 4340 alloy steel during severe milling. It was discovered that the cutting speed and lubricant flow rate significantly affect the amount of surface roughness that is reduced.

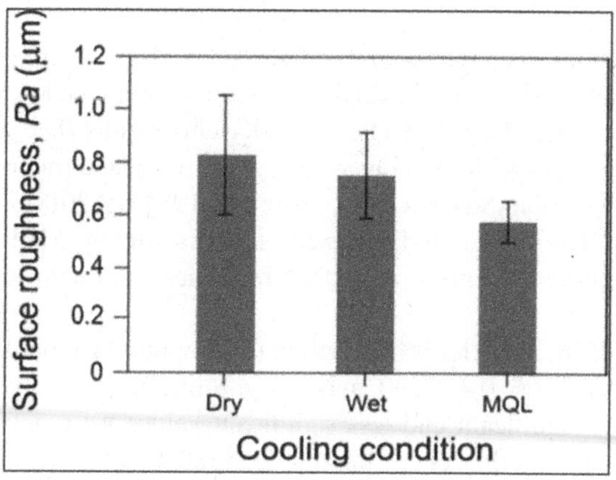

FIGURE 3.11 Illustration of surface roughness of machined substrate at three stages of cooling condition.

Source: Priarone et al. (2014)

Surface roughness was reduced by 16% in MQL with air and 40% in MQL with oxygen, according to a study by Gatade et al. (2016). Analysis shows that cutting at 75 and 100 m/min considerably reduces surface roughness in both air- and oxygen-rich MQL.

In contrast to the inundated supply of lubricants, Kumar et al.'s (2017) MQL approach enhanced the surface quality by 7% to 10%. In their analysis, they discovered that using the MQL process improves the surface polish when likened to traditional wet and dry machining. In contrast to MQL cooling, where both convection and evaporation occur, the authors discover that the flooded cooling system only experiences convection.

When using coated carbide tool inserts to ream aluminium alloy, Lugscheider et al. (1997) examined the impact of the MQL technique over dry machining. It was found that the surface roughness values of machined holes can be reduced with reaming in comparison to dry machining.

Boswell and Islam (2012) examined the impact of the aluminium alloy during end milling. They discovered that mist cooling (air plus lubricant) produced superior surface finishes than dry machining. Additionally, it utilized less lubrication, which is healthier for the environment and people's health.

Lohar and Nanavaty (2013) looked at how turning affected AISI 4330 steel when using dry, wet, and MQL cooling methods. When compared to dry and wet machining, it was discovered that adopting the MQL process improved surface polish by roughly 30%. ANOVA is also used to determine how much the performance elements contribute.

In their study of the impact of the MQL technique on aluminium alloy, Kelly and Cotterell (2002) discovered that in order to increase the material's surface quality, a large amount of coolant or lubricant is needed along with a higher cutting velocity.

Researchers Davim et al. (2007), examined the impact of varying the lubricant flow rate, cutting velocity and feed rate on a brass specimen. The MQL lubrication method was found to yield results in the turning of brass specimens that are on par with those achieved using a wet lubrication system.

While machining steel and Ti-6Al-4V, several authors examined how the MQL method complemented more conventional flood cooling and cryogenic chilling. They have determined that the MQL method yields results that are comparable to flood cooling and cryogenic cooling with regard to surface roughness, microhardness, and chip reduction (Sharma et al., 2016; Joshi & Das, 2018).

Yazid et al. (2019) did a study between dry machining, MQL machining, and cryogenic machining by aiming at surface roughness and chip formation than the latter two.

Çakir et al. (2016) used the MQL method to compute the input factors with lubricating flow rates of 0.25, 0.45, 0.90, and 3.25 ml/min, yielding a wide range of possible results. The significant and less significant values were determined using an ANOVA. They came to the conclusion that feed rate and lubricant flow rate had the biggest impact on surface roughness, whereas cutting velocity had a smaller impact than feed rate and lubricant flow rate. Similar to this, Kouam et al. (2015) looked at how the turning of the aluminium alloy 7075-T6 will be affected by dry and MQL machining. They came to the conclusion that in comparison to dry machining, the MQL process produced better surface finishes and fewer chip formations. Yazid and Zianol (2019) looked into how the MQL parameter and machining parameters affected the milling of an aluminium alloy. They discovered that MQL machining produced comparable outcomes to dry machining.

Tosun and Huseyinoglu (2010) [69] looked and examined the effects of the MQL approach, as opposed to flood cooling, on milling operations with an aluminium alloy (AA-7075). Analysis revealed that MQL cooling yielded a higher-quality surface finish than flood cooling. Senevirathne and Punchihewa (2017) examined the effectiveness of MQL in comparison to flood and dry cooling systems on various steel specimens. Following analysis, it was discovered that MQL technology, together with surface roughness during milling, performs better than flood and dry cooling systems.

To this end, Conger et al. (2019) compared MQL milling operations on the aluminium alloy 6061 to dry machining using two spray nozzles at varying feed, speed, and coolant flow rates. After analysis, it was discovered that the MQL cooling approach created a superior surface finish than dry machining while simultaneously saving the environment and lubricant costs.

The MQL technique may be applied to flood machining without significantly influencing the machining outcomes, according to the authors' research (Fratila & Caizar, 2011). The various factors that influence the required machined characteristics at the time of the milling process are depicted in Figure 3.12, according to the literature review.

Metal cutting machining relies on the cutting fluid to remove chips from the heat-affected zone and cool the cutting tool and workpiece surface. Yet there can be serious consequences for human health and the environment if cutting fluid is used

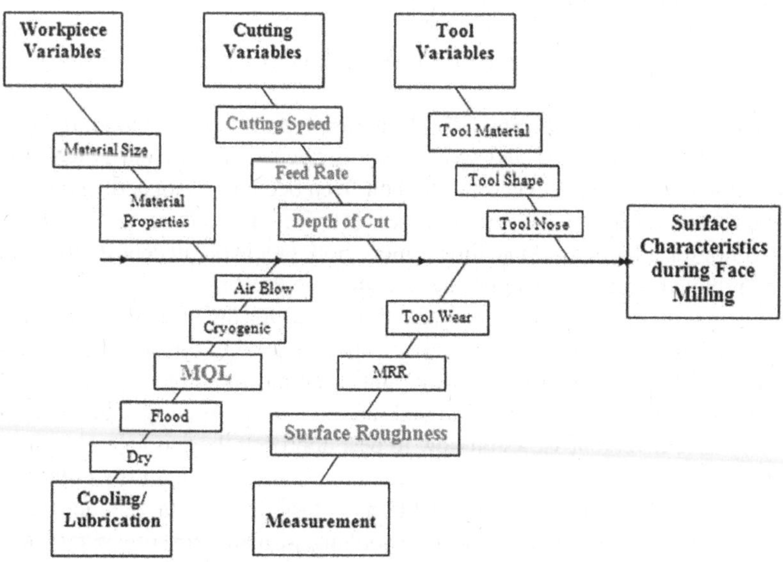

FIGURE 3.12 Surface characteristics during face milling.

incorrectly or disposed of in the wrong way. The majority of experiments have demonstrated that MQL application results in a superior surface to flood and dry machining. Cutting speed, depth of cut, feed rate, and tool nose radius all play significant roles in determining the final surface polish during a turning operation. Cutting zone temperatures rise naturally when steel is turned at high speeds. Cutting tools deform and break under the strain of temperatures so high, micro cracks form on surfaces and in the depths of materials, rust sets in, etc. This research proves that nano fluid MQL may replace flood lubrication to produce the same high-quality surface (Patole et al., 2021).

Nowadays, various additive and subtractive manufacturing processes are going forward to improve several machining characteristics (Kumar et al., 2021, 2019, 2018; Pathri et al., 2023; Babbar et. al., 2022a, 2022b, 2022c, 2022d, 2021a, 2021b, 2021c, 2021d, 2021e, 2021f, 2020a, 2020b, 2020c, 2020d, 2020e, 2019a, 2019b, 2019c, 2017; Kalia et al., 2022; Prakash et al., 2021; Singh et al., 2021c, 2020b; Baraiya et al., 2020; Sharma et al., 2022a, 2022b, 2022c, 2022d, 2022e, 2021a, 2021b, 2021c, 2020a, 2019a, 2019b, 2019c, 2018a, 2018b).

In ultrasonic machining (USM), a rotating diamond-coated tool or floating abrasive particles in a fluid do the actual cutting. The machine's cutting tool is comprised of a relatively soft substance compared to the workpiece. The tool is typically made out of nickel and soft steel. Vibrations of the tool introduce a liquid containing abrasive grains and particles called slurry. The workpieces are immersed in the abrasive slurry until they contact with the grains. The frequency of the vibrating tool determines how long an ultrasonic machine is used. The abrasive slurry's particle size, rigidity, and viscosity all play a role. Boron carbide and silicon carbide grains are used in the abrasive fluid because they are stiffer than others. If the viscosity of the slurry fluid is low enough, the abrasive can be easily removed. Hammering, impact, and cavitation are often cited as the primary cutting mechanisms of USM. In addition, the horn or sonotrode sets in motion the hammering mechanism, which in turn hammers on the abrasives, causing them to penetrate the work materials.

In rotary ultrasonic machining (RUM), abrasive particles are bound to the tool's surface before it is activated and spun to remove material. Coolant is fed into the drill's core to remove debris, keep the drill bit from becoming clogged, and maintain a constant temperature for the drill and the workpiece. In addition to these benefits, RUM allows for a higher quality finish, more hole precision, deeper drilling, faster material removal, and less tool pressure (Singh et al., 2021b).

These days, many non-traditional methods are used for the machining of materials. There are many opportunities for creative problem-solving during machining processes like drilling, slot cutting, milling, and grinding. Examples of these flaws or difficulties include cracks (both radial and lateral), surface quality, and so on. Most academics have divergent views on how to best approach these issues. Despite their best efforts, scientists and researchers were unable to eradicate it. Further other literature shows the utilization of several conditions of usability of lubricant under several machining processes. Further, the surface roughness, material removal rate, cutting forces, tool wear rate, machining temperature, and other machining parameters were discussed in these studies. Finally, it is stated that adjusting the significant MQL machining factors has enhanced the material removal rate and surface finish, directly benefiting the end uses of the machined parts (Akhai & Rana, 2022; Babbar et al., 2021, 2020a, 2020b; Khanduja et al. 2021; Kumar et al., 2021; Rampal et al., 2021; Rana & Akhai, 2022; Sharma et al., 2020a, 2023a, 2021; Singh, K., 2020, 2022).

Consequently, MQL is making its way toward the advanced method of machining processes. This technique is meant to reduce the amount of cutting fluids and maintain the optimum cutting force that provides the desired surface quality. The major outcome of the study stated that the concentration of nanoparticles in the base fluid has a positive effect on the tribological properties of the fluid. Since the tribological properties of the cutting fluid enhanced with nanoparticles are superior to those of the base fluid. Further other literature shows the utilization of several conditions of usability of lubricant under several machining processes. Further, the surface roughness, material removal rate, cutting forces, tool wear rate, machining temperature, and other machining parameters were discussed in these studies

3.6 CONCLUSION AND FUTURE SCOPE

This review investigation explores the impact of dry, flood, and minimum quantity lubrication (MQL) machining conditions on the machining properties of a wide range of hard and brittle materials. This article explains why the minimum quantity lubrication (MQL) method is so much better. Some key outcomes of the review article are pointed as follows:

- At cutting speeds of 300 m/min, feed rates of 0.1 and 0.2 mm/rev, and a MQL flow rate of 10 ml/hr, the MQL method produced holes of equal or higher quality than those obtained with flood cooling on an aluminium substrate.
- At a flow rate of 40 ml/hr, the least surface roughness that could be achieved through milling was measured to be 0.865 m. By comparing the MQL method to dry milling, the improvements in surface finish values were 8.8% for 20 ml/hr and 22.5% for 40 ml/hr.
- In another experimental turning study of the aluminium alloy, it was reported that in comparison to dry machining, the MQL process produced better surface finishes and fewer chip formations.
- Using Taguchi and Grey relational analysis techniques, MQL parameters have been optimized during the turning process. Here, vegetable oil cutting fluid is used in experiments to reduce tool wear and surface roughness. It has been determined that a fluid flow rate of 180 ml/hr and a cutting speed of 30 m/min are the ideal values for optimization.
- It shows that the MQL technique would be a cutting-edge machining technique that could be explored further and address more pioneer topics, such as tool wear rate, material removal rate, and deep grooves machining.

REFERENCES

Akhai, S., & Rana, M. (2022). Taguchi-based grey relational analysis of abrasive water jet machining of Al-6061. *Materials Today: Proceedings*, *65*, 3165–3169. https://doi.org/10.1016/j.matpr.2022.05.361

Al-Attar, E. A., Katamish, H. A., & Sallam, H. I. (2013). Effect of cutting tools and luting cement on marginal adaptation of all ceramic restoration. *Dental Journal, 59*(947), 960.

Amrita, M., Srikant, R. R., & Sitaramaraju, A. V. (2014). Performance evaluation of nano graphite-based cutting fluid in machining process. *Materials and Manufacturing Processes, 29*(5), 600–605. https://doi.org/10.1080/10426914.2014.893060

Attanasio, A., Gelfi, M., Giardini, C., & Remino, C. A. R. L. O. (2006). Minimal quantity lubrication in turning: Effect on tool wear. *Wear, 260*(3), 333–338. https://doi.org/10.1016/j.wear.2005.04.024

Babbar, A., Jain, V., & Gupta, D. (2019a). Thermogenesis mitigation using ultrasonic actuation during bone grinding: A hybrid approach using CEM43° C and Arrhenius model. *Journal of the Brazilian Society of Mechanical Sciences and Engineering, 41*, 1–14. https://doi.org/10.1007/s40430-019-1913-6

Babbar, A., Jain, V., & Gupta, D. (2019d). Neurosurgical bone grinding. *Biomanufacturing*, 137–155. https://doi.org/10.1007/978-3-030-13951-3_7

Babbar, A., Jain, V., & Gupta, D. (2020d). In vivo evaluation of machining forces, torque, and bone quality during skull bone grinding. *Proceedings of the Institution of Mechanical Engineers, Part H: Journal of Engineering in Medicine, 234*(6), 626–638. https://doi.org/10.1177/0954411920911499

Babbar, A., Jain, V., Gupta, D., & Agrawal, D. (2021b). Histological evaluation of thermal damage to Osteocytes: A comparative study of conventional and ultrasonic-assisted bone grinding. *Medical Engineering & Physics, 90*, 1–8. https://doi.org/10.1016/j.medengphy.2021.01.009

Babbar, A., Jain, V., Gupta, D., & Agrawal, D. (2021f). Finite element simulation and integration of CEM43° C and Arrhenius models for ultrasonic-assisted skull bone grinding: A thermal dose model. *Medical Engineering & Physics, 90*, 9–22. https://doi.org/10.1016/j.medengphy.2021.01.008

Babbar, A., Jain, V., Gupta, D., Agrawal, D., & Prakash, C. (2021d). Experimental investigation and parametric optimization of neurosurgical bone grinding under biomimic environment. *Surface Review and Letters*, 2141005. https://doi.org/10.1142/S0218625X21410055

Babbar, A., Jain, V., Gupta, D., Agrawal, D., Prakash, C., Singh, S., . . . & Bogdan-Chudy, M. (2021c). Experimental analysis of wear and multi-shape burr loading during neurosurgical bone grinding. *Journal of Materials Research and Technology, 12*, 15–28. https://doi.org/10.1016/j.jmrt.2021.02.060

Babbar, A., Jain, V., Gupta, D., Prakash, C., & Agrawal, D. (2022b). Potential application of CEM43° C and Arrhenius model in neurosurgical bone grinding. In *Numerical Modelling and Optimization in Advanced Manufacturing Processes* (pp. 145–158). Cham: Springer International Publishing. https://doi.org/10.1007/978-3-031-04301-7_9

Babbar, A., Jain, V., Gupta, D., Prakash, C., & Sharma, A. (2020c). Fabrication and machining methods of composites for aerospace applications. In *Characterization, Testing, Measurement, and Metrology* (1st ed., pp. 109–124). Boca Raton, FL: CRC Press.

Babbar, A., Jain, V., Gupta, D., Prakash, C., & Sharma, A. (2020e). Fabrication and machining methods of composites for aerospace applications. In *Characterization, Testing, Measurement, and Metrology* (pp. 109–124). Boca Raton, FL: CRC Press.

Babbar, A., Jain, V., Gupta, D., Prakash, C., Singh, S., & Sharma, A. (2020b). 3D bioprinting in pharmaceuticals, medicine, and tissue engineering applications. In *Advanced Manufacturing and Processing Technology* (1st ed., pp. 147–161). Boca Raton, FL: CRC Press.

Babbar, A., Jain, V., Gupta, D., Prakash, C., Singh, S., & Sharma, A. (2020f). Effect of process parameters on cutting forces and osteonecrosis for orthopedic bone drilling applications. In *Characterization, Testing, Measurement, and Metrology* (pp. 93–108). Boca Raton, FL: CRC Press.

Babbar, A., Jain, V., Gupta, D., & Sharma, A. (2020a). Fabrication of microchannels using conventional and hybrid machining processes. In *Non-Conventional Hybrid Machining Processes* (1st ed., pp. 37–51). Boca Raton, FL: CRC Press.

Babbar, A., Jain, V., Gupta, D., & Sharma, A. (2020g). Fabrication of microchannels using conventional and hybrid machining processes. *Non-Conventional Hybrid Machining Processes*, 37–51.

Babbar, A., Kumar, A., Jain, V., & Gupta, D. (2019b). Enhancement of activated tungsten inert gas (A-TIG) welding using multi-component TiO2-SiO2-Al2O3 hybrid flux. *Measurement, 148*, 106912. https://doi.org/10.1016/j.measurement.2019.106912

Babbar, A., Rai, A., & Sharma, A. (2021a). Latest trend in building construction: Three-dimensional printing. *Journal of Physics: Conference Series, 1950*(2021), 012007.

Babbar, A., Sharma, A., Jain, V., & Gupta, D. (Eds.). (2022a). *Additive Manufacturing Processes in Biomedical Engineering: Advanced Fabrication Methods and Rapid Tooling Techniques*. Boca Raton, FL: CRC Press.

Babbar, A., Sharma, A., Jain, V., & Jain, A. K. (2019c). Rotary ultrasonic milling of C/SiC composites fabricated using chemical vapor infiltration and needling technique. *Materials Research Express*, *6*(8), 085607. https://doi.org/10.1088/2053-1591/ab1bf7

Babbar, A., Sharma, A., Kumar, R., Pundir, P., & Dhiman, V. (2021e). Functionalized biomaterials for 3D printing: An overview of the literature. *Additive Manufacturing with Functionalized Nanomaterials*, 87–107. https://doi.org/10.1016/B978-0-12-823152-4.00005-3

Babbar, A., Sharma, A., & Singh, P. (2022c). Multi-objective optimization of magnetic abrasive finishing using grey relational analysis. *Materials Today: Proceedings*, *50*, 570–575. https://doi.org/10.1016/j.matpr.2021.01.004

Babbar, A., Singh, P., & Farwaha, H. S. (2017). Parametric study of magnetic abrasive finishing of UNS C26000 flat brass plate. *The International Journal of Advanced Mechatronics and Robotics*, *9*, 83–89.

Baraiya, R., Babbar, A., Jain, V., & Gupta, D. (2020). In-situ simultaneous surface finishing using abrasive flow machining via novel fixture. *Journal of Manufacturing Processes*, *50*, 266–278. https://doi.org/10.1016/j.jmapro.2019.12.051

Behera, B. C., Chetan, S. G., & Rao, P. V. (2014, December). Effects on forces and surface roughness during machining Inconel 718 alloy using minimum quantity lubrication. In *Fifth International and 26th All India Manufacturing Technology, Design and Research Conference (AIMTDR), Assam, India, Dec* (pp. 12–14). https://www.researchgate.net/publication/270648582_EFFECTS_ON_FORCES_AND_SURFACE_ROUGHNESS_DURING_MACHINING_INCONEL_718_ALLOY_USING_MINIMUM_QUANTITY_LUBRICATION

Bhowmick, S., & Alpas, A. T. (2011). The role of diamond-like carbon coated drills on minimum quantity lubrication drilling of magnesium alloys. *Surface and Coatings Technology*, *205*(23–24), 5302–5311. https://doi.org/10.1016/j.surfcoat.2011.05.037

Boswell, B., & Islam, M. N. (2012). Feasibility study of adopting minimal quantities of lubrication for end milling aluminium. *Lecture Notes in Engineering and Computer Science*, 1358–1362. http://hdl.handle.net/20.500.11937/20646

Braga, D. U., Diniz, A. E., Miranda, G. W., & Coppini, N. L. (2002). Using a minimum quantity of lubricant (MQL) and a diamond coated tool in the drilling of aluminum—silicon alloys. *Journal of Materials Processing Technology*, *122*(1), 127–138. https://doi.org/10.1016/S0924-0136(01)01249-3

Brinksmeier, E., Walter, A., Janssen, R., & Diersen, P. (1999). Aspects of cooling lubrication reduction in machining advanced materials. *Proceedings of the Institution of Mechanical Engineers, Part B: Journal of Engineering Manufacture*, *213*(8), 769–778. https://doi.org/10.1243/0954405991517209

Bruni, C., d'Apolito, L., Forcellese, A., Gabrielli, F., & Simoncini, M. (2008). Surface roughness modelling in finish face milling under MQL and dry cutting conditions. *International Journal of Material Forming*, *1*, 503–506. https://doi.org/10.1007/s12289-008-0151-8

Bruni, C., Forcellese, A., Gabrielli, F., & Simoncini, M. (2006). Effect of the lubrication-cooling technique, insert technology and machine bed material on the workpart surface finish and tool wear in finish turning of AISI 420B. *International Journal of Machine Tools and Manufacture*, *46*(12–13), 1547–1554. https://doi.org/10.1016/j.ijmachtools.2005.09.007

Byrne, G., & Scholta, E. (1993). Environmentally clean machining processes—a strategic approach. *CIRP Annals*, *42*(1), 471–474. https://doi.org/10.1016/S0007-8506(07)62488-3

Çakır, A., Yağmur, S. E. L. Ç. U. K., Kavak, N., Küçüktürk, G. Ö. K. H. A. N., & Şeker, U. (2016). The effect of minimum quantity lubrication under different parameters in the turning of AA7075 and AA2024 aluminium alloys. *The International Journal of Advanced Manufacturing Technology*, *84*, 2515–2521. https://doi.org/10.1007/s00170-015-7878-4

Chakraborty, P., Asfour, S., Cho, S., Onar, A., & Lynn, M. (2008). Modeling tool wear progression by using mixed effects modeling technique when end-milling AISI 4340 steel. *Journal of Materials Processing Technology*, *205*(1–3), 190–202. https://doi.org/10.1016/j.jmatprotec.2007.11.197

Conger, D. B., Emiroglu, U., & Altan, E. (2019). An experimental study on cutting forces and surface roughness in MQL milling of aluminum 6061. *Machines Technologies Materials*, *13*(2), 86–89.

Davim, J. P., Sreejith, P. S., & Silva, J. (2007). Turning of brasses using minimum quantity of lubricant (MQL) and flooded lubricant conditions. *Materials and Manufacturing Processes*, *22*(1), 45–50. https://doi.org/10.1080/10426910601015881

Fratila, D., & Caizar, C. (2011). Application of Taguchi method to selection of optimal lubrication and cutting conditions in face milling of AlMg3. *Journal of Cleaner Production*, *19*(6–7), 640–645. https://doi.org/10.1016/j.jclepro.2010.12.007

Gatade, V. T., Patil, V. T., Kuppan, P., Balan, A. S. S., & Oyyaravelu, R. (2016, September). Experimental investigation of machining parameter under MQL milling of SS304. In *IOP Conference Series: Materials Science and Engineering* (Vol. 149, No. 1, p. 012023). IOP Publishing. https://doi.org/10.1088/1757-899X/149/1/012023

Greaves, I. A., Eisen, E. A., Smith, T. J., Pothier, L. J., Kriebel, D., Woskie, S. R., . . . & Monson, R. R. (1997). Respiratory health of automobile workers exposed to metal-working fluid aerosols: Respiratory symptoms. *American Journal of Industrial Medicine*, *32*(5), 450–459. https://doi.org/10.1002/(SICI)1097-0274(199711)32:5<450::AID-AJIM4>3.0.CO;2-W

Hadad, M., & Sadeghi, B. (2013). Minimum quantity lubrication-MQL turning of AISI 4140 steel alloy. *Journal of Cleaner Production*, *54*, 332–343. https://doi.org/10.1016/j.jclepro.2013.05.011

Hassanpour, H., Sadeghi, M. H., Rasti, A., & Shajari, S. (2016). Investigation of surface roughness, microhardness and white layer thickness in hard milling of AISI 4340 using minimum quantity lubrication. *Journal of Cleaner Production*, *120*, 124–134. https://doi.org/10.1016/j.jclepro.2015.12.091

Heisel, U., Lutz, M., Spath, D., Wassmer, R., & Walter, U. (1998). *The Minimum Quantity of Fluid Technique and Its Application to Metal Cutting* (pp. 22–38). Brazil, Maquinas e Metais, Aranda Publisher.

Howes, T. D., Toenshoff, H. K., Heuer, W., & Howes, T. (1991). Environmental aspects of grinding fluids. *CIRP Annals*, *40*(2), 623–630. https://doi.org/10.1016/S0007-8506(07)61138-X

Hwang, Y. K., & Lee, C. M. (2010). Surface roughness and cutting force prediction in MQL and wet turning process of AISI 1045 using design of experiments. *Journal of Mechanical Science and Technology*, *24*, 1669–1677. https://doi.org/10.1007/s12206-010-0522-1

Joshi, K. K., & Das, R. K. (2018, July). Analysis of chip reduction coefficient in turning of Ti-6Al-4V ELI. In *IOP Conference Series: Materials Science and Engineering* (Vol. 390, No. 1, p. 012113). IOP Publishing. https://doi.org/10.1088/1757-899X/390/1/012113

Kalami, H., & Urbanic, J. (2021). Exploration of surface roughness measurement solutions for additive manufactured components built by multi-axis tool paths. *Additive Manufacturing*, *38*, 101822. https://doi.org/10.1016/j.addma.2020.101822

Kalhofer, E. (1997). Dry machining—principles and applications. In *2nd International Seminar on High Technology*. UNIMEP, SP, Brazil.

Kalia, G., Sharma, A., & Babbar, A. (2022). Use of three-dimensional printing techniques for developing biodegradable applications: A review investigation. *Materials Today: Proceedings*. https://doi.org/10.1016/j.matpr.2022.03.445

Kelly, J. F., & Cotterell, M. G. (2002). Minimal lubrication machining of aluminium alloys. *Journal of Materials Processing Technology*, *120*(1–3), 327–334. https://doi.org/10.1016/S0924-0136(01)01126-8

Khan, M. M. A., Mithu, M. A. H., & Dhar, N. R. (2009). Effects of minimum quantity lubrication on turning AISI 9310 alloy steel using vegetable oil-based cutting fluid. *Journal of Materials Processing Technology*, *209*(15–16), 5573–5583. https://doi.org/10.1016/j.jmatprotec.2009.05.014

Khanduja, P., Bhargave, H., Babbar, A., Pundir, P., & Sharma, A. (2021). Development of two-dimensional plotter using programmable logic controller and human machine interface. *Journal of Physics: Conference Series*, *1950*(2021), 012012.

Klocke, F. A. E. G., & Eisenblätter, G. (1997). Dry cutting. *CIRP Annals*, *46*(2), 519–526. https://doi.org/10.1016/S0007-8506(07)60877-4

Kouam, J., Songmene, V., Balazinski, M., & Hendrick, P. (2015). Effects of minimum quantity lubricating (MQL) conditions on machining of 7075-T6 aluminum alloy. *The International Journal of Advanced Manufacturing Technology*, *79*, 1325–1334. https://doi.org/10.1007/s00170-015-6940-6

Kumar, M., Babbar, A., Sharma, A., & Shahi, A. S. (2019, July). Effect of post weld thermal aging (PWTA) sensitization on micro-hardness and corrosion behavior of AISI 304 weld joints. *Journal of Physics: Conference Series*, *1240*(1), 012078. IOP Publishing. https://doi.org/10.1088/1742-6596/1240/1/012078

Kumar, M., Sharma, A., & Shahi, A. S. (2019). A sensitization studies on the metallurgical and corrosion behavior of AISI 304 SS welds. In *Advances in Manufacturing Processes: Select Proceedings of ICEMMM 2018* (pp. 257–265). Singapore: Springer. https://doi.org/10.1007/978-981-13-1724-8_25

Kumar, S., Singh, D., & Kalsi, N. S. (2017). Analysis of surface roughness during machining of hardened AISI 4340 steel using minimum quantity lubrication. *Materials Today: Proceedings*, *4*(2), 3627–3635. https://doi.org/10.1016/j.matpr.2017.02.255

Kumar, S., Sudhakar, R. P., Goyal, D., & Sehgal, S. (2021). Process modelling for machining Inconel 825 using cryogenically treated carbide insert. *Metal Powder Report*, *76*, 66–74. https://doi.org/10.1016/j.mprp.2020.06.001

Kumar, V., Prakash, C., Babbar, A., Choudhary, S., Sharma, A., & Uppal, A. S. (2017). Additive manufacturing in biomedical engineering: Present and future applications. In *Additive Manufacturing Processes in Biomedical Engineering* (pp. 143–164). Boca Raton, FL: CRC Press. https://doi.org/10.1201/9781003217961-8

Lawal, S. A., Choudhury, I. A., & Nukman, Y. (2013). A critical assessment of lubrication techniques in machining processes: A case for minimum quantity lubrication using vegetable oil-based lubricant. *Journal of Cleaner Production*, *41*, 210–221. https://doi.org/10.1016/j.jclepro.2012.10.016

Li, K. M., & Chou, S. Y. (2010). Experimental evaluation of minimum quantity lubrication in near micro-milling. *Journal of Materials Processing Technology*, *210*(15), 2163–2170. https://doi.org/10.1016/j.jmatprotec.2010.07.031

Lohar, D. V., & Nanavaty, C. R. (2013). Performance evaluation of minimum quantity lubrication (MQL) Using CBN tool during hard turning of AISI 4340 and its comparison with dry and wet turning. *Bonfring International Journal of Industrial Engineering and Management Science*, *3*(3), 102–106. https://doi.org/10.9756/BIJIEMS.4392

Lugscheider, E., Knotek, O., Barimani, C., Leyendecker, T., Lemmer, O., & Wenke, R. (1997). Investigations on hard coated reamers in different lubricant free cutting operations. *Surface and Coatings Technology*, *90*(1–2), 172–177. https://doi.org/10.1016/S0257-8972(96)03114-3

Marksberry, P. W., & Jawahir, I. S. (2008). A comprehensive tool-wear/tool-life performance model in the evaluation of NDM (near dry machining) for sustainable manufacturing. *International Journal of Machine Tools and Manufacture*, *48*(7–8), 878–886. https://doi.org/10.1016/j.ijmachtools.2007.11.006

Nasir, A. (1998). General comments on ecological and dry machining. In *Network Proceedings "Technical Solutions to Decrease Consumption of Cutting Fluids," Sobotin-Sumperk, Czech Republic* (pp. 10–14). https://astakhov.tripod.com/MC/Proceedings-EcologEU.pdf#page=10

Pathri, B. P., Khan, M. S., & Babbar, A. (2023). Relevance of Bio-Inks for 3D bioprinting. In *Additive Manufacturing Processes in Biomedical Engineering* (pp. 81–98). Boca Raton, FL: CRC Press.

Patole, P. B., Kulkarni, V. V., & Bhatwadekar, S. G. (2021). MQL Machining with nano fluid: A review. *Manufacturing Review*, *8*, 13. https://doi.org/10.1051/mfreview/2021011

Prakash, C., Kumar, V., Mistri, A., Uppal, A. S., Babbar, A., Pathri, B. P., & Zheng, H. (2021). Investigation of functionally graded adherents on failure of socket joint of FRP composite tubes. *Materials*, *14*(21), 6365. https://doi.org/10.3390/ma14216365

Priarone, P. C., Robiglio, M., Settineri, L., & Tebaldo, V. (2014). Milling and turning of titanium aluminides by using minimum quantity lubrication. *Procedia Cirp*, *24*, 62–67. https://doi.org/10.1016/j.procir.2014.07.147

Rahman, M., Kumar, A. S., & Salam, M. U. (2002). Experimental evaluation on the effect of minimal quantities of lubricant in milling. *International Journal of Machine Tools and Manufacture*, *42*(5), 539–547. https://doi.org/10.1016/S0890-6955(01)00160-2

Rampal, R., Goyal, T., Goyal, D., Mittal, M., Dang, R. K., & Bahl, S. (2021). Magnetorheological abrasive finishing (MAF) of soft material using abrasives. *Materials Today: Proceedings*, *45*, 5114–5121. https://doi.org/10.1016/j.matpr.2021.01.629

Rana, M., & Akhai, S. (2022). Multi-objective optimization of Abrasive water jet Machining parameters for Inconel 625 alloy using TGRA. *Materials Today: Proceedings*, *65*, 3205–3210. https://doi.org/10.1016/j.matpr.2022.05.374

Sahoo, P., Barman, T. K., & Routara, B. C. (2008). Taguchi based fractal dimension modelling and optimization in CNC turning. *Advances in Production Engineering & Management*, *3*(4), 205–217.

Sales, W. F., Diniz, A. E., & Machado, Á. R. (2001). Application of cutting fluids in machining processes. *Journal of the Brazilian Society of Mechanical Sciences*, *23*, 227–240. https://doi.org/10.1590/S0100-73862001000200009

Sanchez, J. A., Pombo, I., Alberdi, R., Izquierdo, B., Ortega, N., Plaza, S., & Martinez-Toledano, J. (2010). Machining evaluation of a hybrid MQL-CO2 grinding technology. *Journal of Cleaner Production*, *18*(18), 1840–1849. https://doi.org/10.1016/j.jclepro.2010.07.002

Sarıkaya, M., & Güllü, A. (2015). Multi-response optimization of minimum quantity lubrication parameters using Taguchi-based grey relational analysis in turning of difficult-to-cut alloy Haynes 25. *Journal of Cleaner Production*, *91*, 347–357. https://doi.org/10.1016/j.jclepro.2014.12.020

Senevirathne, S. W. M. A. I., & Punchihewa, H. K. G. (2017, September). Comparison of tool life and surface roughness with MQL, flood cooling, and dry cutting conditions with P20 and D2 steel. In *IOP Conference Series: Materials Science and Engineering* (Vol. 244, No. 1, p. 012006). IOP Publishing. https://doi.org/10.1088/1757-899X/244/1/012006

Sharma, A., Babbar, A., Jain, V., & Gupta, D. (2018a). Enhancement of surface roughness for brittle material during rotary ultrasonic machining. In *MATEC Web of Conferences* (Vol. 249, p. 01006). EDP Sciences. https://doi.org/10.1051/matecconf/201824901006

Sharma, A., Babbar, A., Tian, Y., Pathri, B. P., Gupta, M., & Singh, R. (2022b). Machining of ceramic materials: A state-of-the-art review. *International Journal on Interactive Design and Manufacturing (IJIDeM)*, 1–21. https://doi.org/10.1007/s12008-022-01016-7

Sharma, A., Fidan, I., Huseynov, O., Ali, M. A., Alkunte, S., Rajeshirke, M., & . . . Popov, V. (2023a). Recent inventions in additive manufacturing: Holistic review. *Inventions*, *8*(4), 103. https://doi.org/10.3390/inventions8040103

Sharma, A., Jain, V., & Gupta, D. (2018b). Characterization of chipping and tool wear during drilling of float glass using rotary ultrasonic machining. *Measurement*, *128*, 254–263. https://doi.org/10.1016/j.measurement.2018.06.040

Sharma, A., Jain, V., & Gupta, D. (2019a). Tool wear analysis while creating blind holes on float glass using conventional drilling: A multi-shaped tools study. In *Advances in Manufacturing Processes: Select Proceedings of ICEMMM 2018* (pp. 175–183). Singapore: Springer. https://doi.org/10.1007/978-981-13-1724-8_17

Sharma, A., Jain, V., & Gupta, D. (2019b). Comparative analysis of chipping mechanics of float glass during rotary ultrasonic drilling and conventional drilling: For multi-shaped tools. *Machining Science and Technology*, *23*(4), 547–568. https://doi.org/10.1080/10910344.2019.1575402

Sharma, A., Jain, V., & Gupta, D. (2019c). Multi-shaped tool wear study during rotary ultrasonic drilling and conventional drilling for amorphous solid. *Proceedings of the Institution of Mechanical Engineers, Part E: Journal of Process Mechanical Engineering*, *233*(3), 551–560. https://doi.org/10.1177/0954408918776724

Sharma, A., Jain, V., & Gupta, D. (2021c). Effect of pre and post tempering on hole quality of float glass specimen: For rotary ultrasonic and conventional drilling. *Silicon*, *13*, 2029–2039. https://doi.org/10.1007/s12633-020-00597-w

Sharma, A., Jain, V., & Gupta, D. (2022c). Mathematical approach on chipping volume estimation generated during rotary ultrasonic drilling for float glass. *Proceedings of the National Academy of Sciences, India Section A: Physical Sciences*, 1–7. https://doi.org/10.1007/s40010-021-00732-1

Sharma, A., Jain, V., Gupta, D., & Babbar, A. (2020a). A review study on miniaturization. In *Advanced Manufacturing and Processing Technology* (1st ed., pp. 111–131). Boca Raton, FL: CRC Press.

Sharma, A., Jain, V., Gupta, D., & Babbar, A. (2020b). A review study on miniaturization: A boon or curse. *Advanced Manufacturing and Processing Technology*, 111–131.

Sharma, A., Kalsia, M., Uppal, A. S., Babbar, A., & Dhawan, V. (2022a). Machining of hard and brittle materials: A comprehensive review. *Materials Today: Proceedings*, *50*, 1048–1052. https://doi.org/10.1016/j.matpr.2021.07.452

Sharma, A., Kumar, V., Babbar, A., Dhawan, V., Kotecha, K., & Prakash, C. (2021a). Experimental investigation and optimization of electric discharge machining process parameters using grey-fuzzy-based hybrid techniques. *Materials*, *14*(19), 5820. https://doi.org/10.3390/ma14195820

Sharma, A., Sandhu, H. S., Goyal, D., Goyal, T., Jarial, S., & Sharda, A. (2023b). Sustainable development in cold gas dynamic spray coating process for biomedical applications: Challenges and future perspective review. *International Journal on Interactive Design and Manufacturing (IJIDeM)*, 1–17. https://doi.org/10.1007/s12008-023-01474-7

Sharma, J., & Sidhu, B. S. (2014). Investigation of effects of dry and near dry machining on AISI D2 steel using vegetable oil. *Journal of Cleaner Production*, *66*, 619–623. https://doi.org/10.1016/j.jclepro.2013.11.042

Sharma, A., Singh, R. K., Dixit, A. R., & Tiwari, A. K. (2016). Characterization and experimental investigation of Al2O3 nanoparticle based cutting fluid in turning of AISI 1040 steel under minimum quantity lubrication (MQL). *Materials Today: Proceedings*, *3*(6), 1899–1906. https://doi.org/10.1016/j.matpr.2016.04.090

Sharma, V. K., Rana, M., Singh, T., Singh, A. K., & Chattopadhyay, K. (2021b). Multi-response optimization of process parameters using Desirability Function Analysis during machining of EN31 steel under different machining environments. *Materials Today: Proceedings*, *44*, 3121–3126. https://doi.org/10.1016/j.matpr.2021.02.809

Sharma, V. K., Singh, T., Singh, K., & Kaur, G. (2022d). MQL assisted face milling of EN-31: Tool wear optimization and its correlation with cutting temperature. *Materials Today: Proceedings*, *71*, 346–351. https://doi.org/10.1016/j.matpr.2022.09.359

Sharma, V. K., Singh, T., Singh, K., Rana, M., & Gehlot, A. (2022e). Optimization of surface qualities in face milling of EN-31 employing hBN nanoparticles-based minimum quantity lubrication. *Materials Today: Proceedings*, *69*, 303–308. https://doi.org/10.1016/j.matpr.2022.08.539

Sharma, V. S., Dogra, M., & Suri, N. M. (2009). Cooling techniques for improved productivity in turning. *International Journal of Machine Tools and Manufacture*, *49*(6), 435–453. https://doi.org/10.1016/j.ijmachtools.2008.12.010

Singh, B. P., Singh, J., Bhayana, M., & Goyal, D. (2021a). Experimental investigation of machining nimonic-80A alloy on wire EDM using response surface methodology. *Metal Powder Report*, *76*, 9–17. https://doi.org/10.1016/j.mprp.2020.12.001

Singh, B. P., Singh, J., Bhayana, M., Singh, K., & Singh, R. (2022). Experimental examination of the machining characteristics of Nimonic 80-A alloy on wire EDM. *Materials Today: Proceedings*, *69*, 291–296. https://doi.org/10.1016/j.matpr.2022.08.537

Singh, G., Babbar, A., Jain, V., & Gupta, D. (2021b). Comparative statement for diametric delamination in drilling of cortical bone with conventional and ultrasonic assisted drilling techniques. *Journal of Orthopaedics*, *25*, 53–58. https://doi.org/10.1016/j.jor.2021.03.017

Singh, J., Singh, C., & Singh, K. (2023a). Rotary ultrasonic machining of advance materials: A review. *Materials Today: Proceedings*. https://doi.org/10.1016/j.matpr.2023.01.159

Singh, K. (2020). *Effect of Minimum Quantity Lubrication Technique on Machining Characteristics of Ductile Material During Face Milling* (Doctoral dissertation, Chitkara University Punjab, India).

Singh, K., Sharma, V. K., Singh, T., Rana, M., Goyal, R., & Rana, A. (2023b). Investigation of Al-6061 alloy in face milling through DFA approach. *Materials Today: Proceedings*. https://doi.org/10.1016/j.matpr.2023.03.031

Singh, K., Sharma, V. K., Singh, T., Singh, A. K., & Singh, R. (2022). Correlating surface roughness with cutting temperature in face milling of Al-6061using CRITIC approach. *Materials Today: Proceedings*, *69*, 333–338. https://doi.org/10.1016/j.matpr.2022.08.545

Singh, K., Singh, A. K., & Chattopadhyay, K. D. (2020a). Selection of optimal cutting conditions and coolant flow rate (CFR) for enhancing surface finish in milling of aluminium alloy. *Materials Today: Proceedings*, *21*, 1520–1524. https://doi.org/10.1016/j.matpr.2019.11.076.

Singh, R., Gupta, R. K., & Tripathi, J. (2014). Surface roughness analysis and compare prediction and experimental value for cylindrical stainless-steel pipe ss 316l in CNC lathe turning process using ANN method for re-optimization and cutting fluid. *International Journal of Engineering Science*, *7*, 58–71.

Singh, S., Prakash, C., Pramanik, A., Basak, A., Shabadi, R., Królczyk, G., . . . & Babbar, A. (2020b). Magneto-rheological fluid assisted abrasive nanofinishing of β-Phase Ti-Nb-Ta-Zr alloy: Parametric appraisal and corrosion analysis. *Materials*, *13*(22), 5156. https://doi.org/10.3390/ma13225156.

Sreejith, P. S., & Ngoi, B. K. A. (2000). Dry machining: Machining of the future. *Journal of Materials Processing Technology*, *101*(1–3), 287–291. https://doi.org/10.1016/S0924-0136(00)00445-3

Tasdelen, B., Wikblom, T., & Ekered, S. (2008). Studies on Minimum Quantity Lubrication (MQL) and air cooling at drilling. *Journal of Materials Processing Technology*, *200*(1–3), 339–346. https://doi.org/10.1016/j.jmatprotec.2007.09.064

Tosun, N., & Huseyinoglu, M. (2010). Effect of MQL on surface roughness in milling of AA7075-T6. *Materials and Manufacturing Processes*, *25*(8), 793–798. https://doi.org/10.1080/10426910903496821

Uysal, A., Demiren, F., & Altan, E. (2015). Applying Minimum Quantity Lubrication (MQL) method on milling of martensitic stainless steel by using nano MoS2 reinforced vegetable cutting fluid. *Procedia-Social and Behavioral Sciences*, *195*, 2742–2747. https://doi.org/10.1016/j.sbspro.2015.06.384

Varadarajan, A. S., Philip, P. K., & Ramamoorthy, B. (2002). Investigations on hard turning with minimal cutting fluid application (HTMF) and its comparison with dry and wet turning. *International Journal of Machine Tools and Manufacture*, *42*, 193–200. https://doi.org/10.1016/S0890-6955(01)00119-5

Weinert, K., Inasaki, I., Sutherland, J. W., & Wakabayashi, T. (2004). Dry machining and minimum quantity lubrication. *CIRP Annals*, *53*(2), 511–537. https://doi.org/10.1016/S0007-8506(07)60027-4

Yazid, M. Z. A., & Zainol, A. (2019). Environmentally friendly approaches assisted machining of aluminum alloy 7075-T6 for automotive applications: A review. *International Journal of Integrated Engineering*, *11*(6), 18–26. https://penerbit.uthm.edu.my/ojs/index.php/ijie/article/view/3556

Yazid, M. Z. A., Zainol, A., & Mustapaha, A. M. (2019). Effect of machining parameters in milling aluminium alloy 7075—T6 under MQL condition. *International Journal of Engineering and Advanced Technology (IJEAT)*, *9*(2), 109–113. https://doi.org/10.35940/ijeat.A22

4 Comprehensive Study on Electrochemical Discharge Machining

Santosh Kumar, Rakesh Kumar, Mohit Kumar

4.1 INTRODUCTION

ECDM is a hybrid accepted worldwide process combining the principles of EDM and ECM. The metal removal completed by spark erosion "rough machining", electrochemical affect and smooth the work surface in this process. The growing demand for micro- and macro-level items and parts of complicated-to-remove material has been rapidly increasing in medical devices (implant and instrument), automobile, optics, aerospace, and electronic industries. Despite their outstanding advantages, lot of these complicated-to-remove material removal appear to have less applications. These metals present numerous threats to traditional machining processes (termed as milling as well as turning). Titanium alloys (Ti-6Al-4V). The low heat conductivity and "excellent chemical reactivity" are outcomes in high cutting temperatures and superior adhesion between the tool and the substrate material, resulting in tool wear (Muniruddin & Ahmed, 2020). The creation of micro components attracts the greatest attention from the industrial industry in order to create little or tiny items that are in high demand in modern society. To meet the demand, scientists and technologists face increasing obstacles as well as issues in the industrial industry. The difficulty in implementing traditional manufacturing methods are caused by three primary factors, including new technologies, materials with high dimensional and precision requirements, and limited machinability economic output rate. The removal of some material is referred to as machining—from a work specimen via direct or indirect interaction with a tool for the development of a specified shape with a predetermined level of precision and surface quality. Parts produced by casting, forming, and different shaping procedures sometimes need additional operations before use or application assembly. Many engineering applications require the interchange of parts in order to work effectively and consistently over their projected service lifetimes. Because of their advantageous properties, sophisticated machining methods are required in modern manufacturing sectors to create goods with advanced materials that are utilized in engineering use, such as ceramics, quartz, alumina, and glass. Abrasive water jet machining (AWJM) method has limited applications because to transverse cutting speed, product quality is not very excellent due to poor surface quality and big space area required for installation, expensive investment,

DOI: 10.1201/9781003327905-4

and maintenance cost. Ultrasonic machining (USM) has certain inherent constraints, including tool wear, a high capital cost, and the possibility of tool bending owing to contraction and vibration. Although laser beam machining (LBM), the production of a very wide unwanted heat impacted zone impairs product quality and necessitates a significant expenditure. ECM necessitates a big financial investment, expert labor, and a broad installation area. The disposal of wasted electrolyte and the influence of stray current are other significant downsides of the ECM process. Again, some disadvantages of electro-discharge machining (EDM) are the difficulty of fabricating diverse forms, high metal time of metal removal necessary to manufacture micro products, and the expensive cost of equipment. Furthermore, ECM and EDM are mostly applicable for electrically conducting materials. As a result, an alternate machining method is being developed to cut materials of non-conducting type, like ceramics and glass, with the least amount of investment. As a result, a unique machining procedure is required, which will be beneficial for manufacturing items made of materials which are non-conductive electrically that can withstand the aforementioned unfavorable impacts of the aforementioned machining procedures. Machining by electrochemical discharge (ECDM) technique may mill electrically non-conductive materials (e.g. ceramics and glass) when compared to the preceding machining methods, heat impacts (i.e. the establishment of a HAZ is minimal). There is no need for an experienced person to operate. The ECDM equipment and procedure are unaffected by the physical as well as chemical characteristics of the material (Kurafuji & Suda, 1968). ECM utilized the removal of metal by anodic dissolution and applicable in aerospace, medical equipment, power supply company and automotive, etc. (Kumar et al., 2018, 2019, 2020a, 2020b; Singh et al., 2020). The major focus of additive and subtractive manufacturing methods is to develop 3D products with superior surface quality and tolerance. Further, today's manufacturing industries are implementing both approaches in significant manners (Akhai & Rana, 2022; Babbar et al., 2020a, 2020b, 2020c, 2020d, 2021, 2022; Kalia et al., 2022; Kumar & Kumar, 2021; Parikh et al., 2023; Prakash et al., 2021; Rampal et al., 2021; Rana & Akhai, 2022; Sharma et al., 2018a, 2018b, 2019a, 2019b, 2019c, 2020a, 2020b, 2021a, 2023a, 2023b; Sharma & Jain, 2020; Sharma et al., 2021d).

4.1.1 History of ECDM

ECDM was first advance in 1968 by Kura Fuji for the removal of material from the glass. It combines two well-known processes, EDM and ECM. In 2005, the researchers R. Wuthrich and V. Fascio were given the name "electrochemical spark machining and electrochemical arc machining" of this process. In the current era, the non-conductive metals demand has increased especially in engineering use owing to high chemical resistance, high heat resistance, and high creep; however, metal removal with greater reliability and high accuracy is a major threat. For the machining process in ECDM, various tools (Cu Ti, SS, HSS) are used. Similarly, various electrolytes (such as NaOH, KOH, and NaNO3) are employed during the process to improve the etching effect (Bhattacharya et al., 1999; Wuthrich & Fascio, 2005; Jain & Adhikary, 2008).

In this chapter review, the history, working principle, distinct process parameters, distinct types of materials used, ECDM variants, and future research possibilities are also described.

4.1.2 Working Principle of ECDM

The ECDM is a non-traditional machining process in which metal is removed from the work surface owing to simulation interaction of electrochemical dissolution and spark discharges. In this machining process, DC power sources are connected with two electrodes called anode and cathode. The NaCl, KOH, and NaOH are used as electrolyte solution, upon terminal which both the electrodes are immerged. The anode (+ve terminal) is connected with auxiliary electrode and cathode (-ve terminal) with tool electrode. The electrolysis takes place between positive and negative when voltage is applied.

As the potential difference increases, so does the density of bubbles development at the cathode. At the cathode, bubbles develop and combine to form a thinner gas film that encircles the tool, electrode. The generated gas layer stops current from flowing via the tool electrode to the electrolyte. As a result, a strong electric field is formed, which causes an electric discharge to form. The bubbles development and spark phenomena (Charak & Jawalka, 2019) are depicted in the Figure 4.1 and Figure 4.2 later.

In spark discharge, a stream of concentrated type electrons from the cathode mainly strikes the workpiece with greater speed, creating the compression waves mechanically. Because the job space is very near to the tool electrode, machining begins from the upper surface of substrate because of melting and vaporization. The sparking occurs when the potential difference is highly more in the film (Ali et al., 2019).

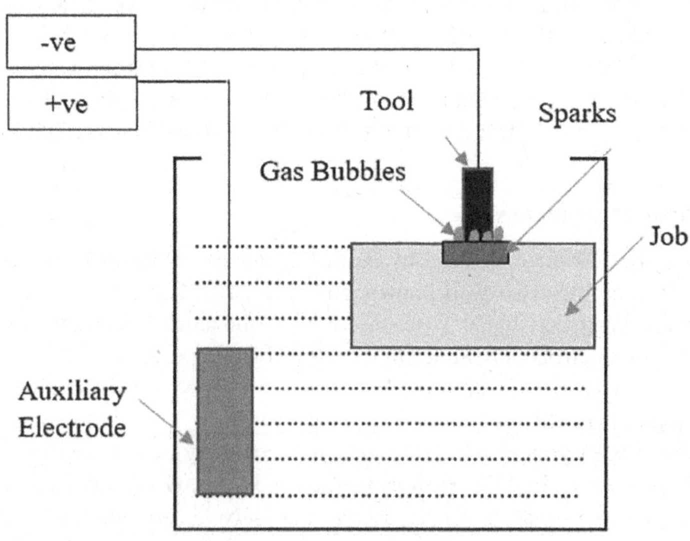

FIGURE 4.1 Bubble development and spark phenomena.

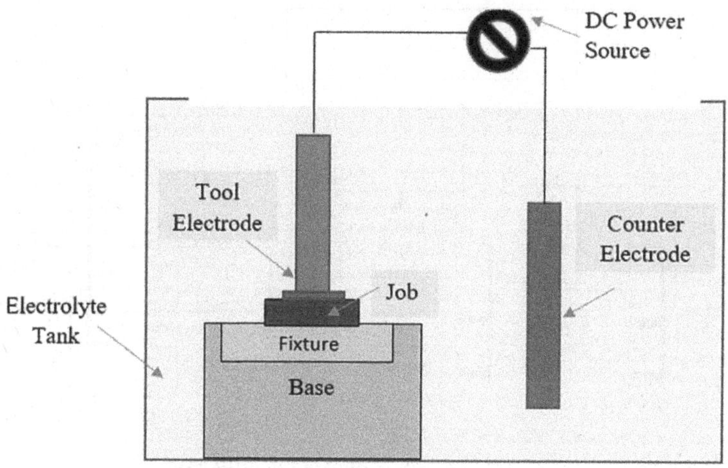

FIGURE 4.2 ECDM process.

4.1.3 Distinct Process Parameters of ECDM

ECDM is considered as complicated process. This machining process mainly depends upon distinct process parameters (e.g. voltage, MRR, electrolyte blend, speed and feed rate, tool shape, duty cycle, tool material, tool wear rate, surface roughness, etc.). The general parameters which affect the performance of the ECDM process are illustrated in Figure 4.3.

However, it is unclear which one major factor has the greatest impact on metal removal performance and how the best parameters can be optimized. The optimal process parameters are critical for improving product quality and process performance. As a result, proper parameter optimization is critical for understanding the interaction between process parameters as well as the response parameters in the ECDM process (Jain & Priyadarshini, 2014). The detailed study of some critical process parameters is explained in the next section.

4.1.4 Applied DC Voltage

A power supply (DC) voltage is provided given the auxiliary electrode and tool electrode. The machining rate enhances as the machining voltage increases. The rate of metal removal mainly reaches a max at a specific voltage and subsequently drops. However, DC source (that maintains a consistent voltage) throughout the process is found to be the highly effective for ECDM (Singh & Singhal, 2016).

4.1.5 Types of Electrolyte

ECDM process generally utilized many types of electrolytes as shown in Figure 4.4. Among all these electrolytes owing to attractive properties, NaOH and KOH are most extensively used. Because it rarely enhances the metal removal as compared to other

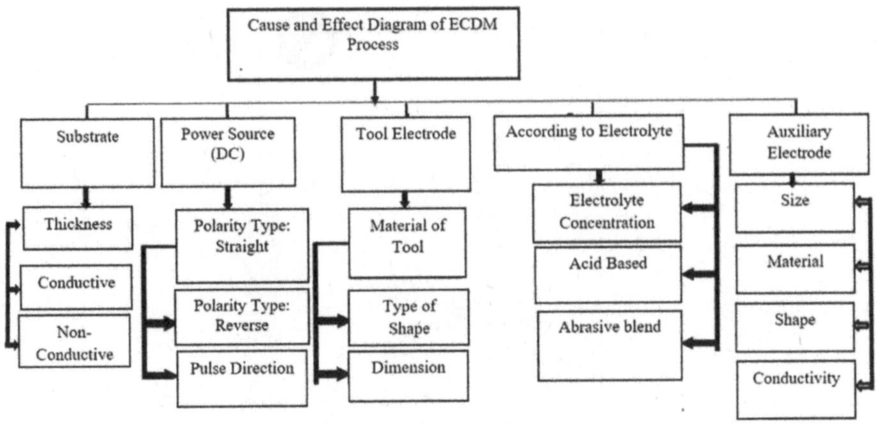

FIGURE 4.3 Flowchart of cause-and-effect illustration of ECDM.

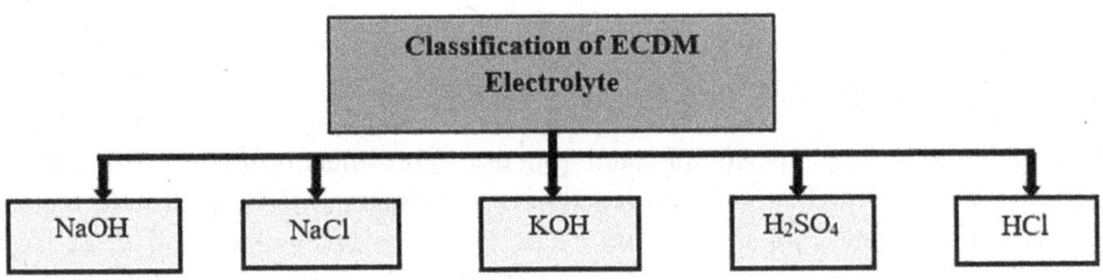

FIGURE 4.4 Distinct types of electrolytes.

electrolytes, the influence of electrolyte on metal removal is difficult to comprehend and cannot be fully stated as a main role of concentration as well as temperature. Because the metal removal process is primarily chemical, the type of the electrolyte has a substantial impact on the machining behavior (Basak & Ghosh, 1996).

4.1.6. Effect of Adding Abrasives

Some researchers observed that an enhancement of surface roughness is achieved by blending the electrolyte with abrasive and by refining the micro-level cracks and heat-affected zone introduced by heat erosion at the time of discharge. Min Seop et al. investigate that increasing the abrasive concentration decreases the surface roughness and improves the MRR (Yang et al., 2006).

4.1.7. Inter-electrode Gap

When the cathode (tool electrode) and anode are detached by a gap (auxiliary electrode). This space is measured in centimeters. Inter-electrode gap mainly affects the MRR. As the gap increases between the electrodes, the material removal rate decreases. The inter-electrode space is a highly critical process parameter, and it should be controlled when the metal is removed (Bhattacharyya et al., 2004; Wang et al., 2007; Neto & Cirilo, 2011; Ozkeskin, 2008; Lu et al., 2011).

4.2 DISTINCT TYPES OF MATERIALS USED IN ECDM

The ECDM process detailed explanation and distinct types of material used are shown in Figure 4.5.

4.2.1 Glass, Ceramics

Glass has several features such as chemical resistance and transparency. It is most widely used in the sector of "lab-on-a-chip" devices as well as micro electromechanical systems. Han et al., proved that machining processes enhance the machining rate by ultrasonic vibrations to electrolyte. Furthermore, the experiment's side-insulated tool electrode reduces overcut (Min-Seop & Byung-Kwon, 2009). The ECDM technology is effective for micro-level machining ceramics. Furthermore, numerous researchers working on the machining of aluminium oxide (Al2O3) ceramics, such as the controlled spark phenomena with changing in tool electrode tip and micro slicing, observed a tool wear mechanism (Sarkar et al., 2006; Singh et al., 1996; Bhattacharya et al., 1999; Chak & Rao, 2008; Dhanvijay & Ahuja, 2014; Abou & Wuthrich, 2012; Manna & Kundal, 2015; Behroozfar & Razfar, 2016).

4.2.2 Super Alloy

The Ni-based super alloy is mostly utilized in the aircraft heart (engine) owing to their excellent characteristics (e.g. heat resistance property). Thus, a new hybrid method, called tube electrode high-speed ECD drilling, is most widely utilized for metal removal of film cooling holes in case of super-alloy materials. Zhang et al., investigated the effects of tubular electrode inner structure on metal removal of super alloys using TEHECDD. The results showed that the double-hole tubular electrode is the best tool electrode among various geometries for increasing machining rate and improving surface quality. Yan et al., performed a comparative investigation of super-alloy metal removal using TEHECDD and EDM. It demonstrated that TEHECDD is encouraging machining process for micro-hole material removal in case of super alloys (Akhtar et al., 2014; Yan et al., 2016a, 2016b; Yadav & Yadava, 2017a; Zhang et al., 2016).

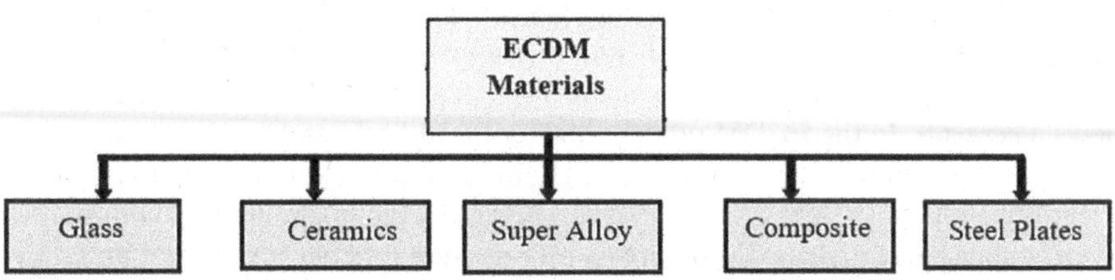

FIGURE 4.5 Electrochemical materials.

4.2.3 Composites

ECDM's process capability allowed it to be deposited to materials of conductive type (steel, MMC, and super alloys). Hofy et al., reported on steel machining with wire electrochemical arc machining in 1988. The composites have several attractive properties such as resistance against wear, high temperature, etc.; hence, they are used in distinct sectors (aerospace, automotive, additive manufacturing [orthopedic], electronics, etc.). Apart from this, the excellent properties of this material make them complex to machine by traditional machining methods. Hence, several studies of researchers proved that metal removal of distinct composites are difficult to cut by non-traditional machining (ultrasonic, electro-discharge machining, LASER, and ECDM method) (Hofy et al., 1998; Speer & Es-Said, 2001; Kunze & Bampton, 2001; Singh et al., 2013a; Singh & Dvivedi, 2016; Zweben, 2005; Antil et al., 2018a, 2018b; Tandon et al., 1990; Kumar & Singh, 2017; Yadav & Yadava, 2017b, Yuan et al., 2017; Taweeporn et al., 2015; Manna & Narang, 2012; Malik & Manna, 2016; Liu et al., 2010).

4.2.4 Steel Plates

Many of the researchers' studies revealed the axial tool wear rate in case of machining of hole of small size on steel plate and HAZ mechanism. Similarly, in another study, Krotz et al., used steel plate that investigates tool wear of conductive material by spark-assisted electrochemical machining. Coteata et al., also studied the drilling speed for materials that have small holes on plate. In the results, the authors explain the consequences of distinct parameters like electrode diameter, voltage, etc. (Coteata et al., 2008, 2009, 2011; Huang et al., 2014; Chavoshi & Behagh, 2014; Krotz & Wegener, 2015).

4.3 ECDM VARIANTS

The principle of ECDM process is to perform distinct operations (milling, drilling, cutting, dressing, turning, and die-sinking). These machining process activities are most effective to develop profile on brittle and hard materials. Some examples of distinct variants in detail are explained next.

4.3.1 Electrochemical Discharge Milling

This metal removal method is employed for developing three-dimensional, complicated microstructures on quartz and glass materials. Many researchers studied surface texturing of micro channels and micro grooves. In this method, a wheel of cylindrical rotating type is used as a cutting tool (cathode electrode). This tool follows a set path. In the ECDM process, the tool travel rate and tool rotation rate are mentioned as main process parameters. Higher tool rotation rates help to eliminate electrolyte replenishment concerns while also producing micro grooves with smaller widths and sharper edges. Surprisingly, tool rotation rate has little effect on groove depth. Because of the sudden tool travel rate, micro grooves shallower type with higher breadth may be produced (Zheng & Cheng, 2007).

4.3.2 Electrochemical Discharge Drilling

Precision holes on thin as well as on thick surfaces according to the demand of high aspect ratio. To satisfy the challenges of micro manufacturing, several researchers employed the ECDD technique to drill through and blind tiny holes. Previously, this technology was utilized to drill materials of conductive type (Cr, Ti, nimonic alloys, cobalt, low-alloy steels). The machined surfaces of these materials have a smooth surface quality, comparable to electrochemically metal removal surfaces. Thus, on these mentioned conductive materials, the ECDD technique has utilized steels, borosilicate glass, e-grassfire-epoxy composites, soda-lime glass, etc. (Silva & McGeough, 1986; West & Jadhav, 2007; Coteaţă et al., 2008a; Ziki & Wüthrich, 2013; Gupta et al., 2014; Paul & Hiremath, 2013; Liao & Peng, 2006; Manna et al., 2012; Sarkar et al., 2006).

4.3.3 Electrochemical Discharge Turning

ECDM with regular rotation of job is a versatile realization of this process to remove the material of circular parts. The presentation of ECDT is depicted in Figure 4.6.

It is made up of a rotating workpiece submerged in an electrolytic solution bath. During machining, the rotating movement of the job allows new electrolyte to be fed over the tiny space between tool and job. The job rotation rate is a major parameter of process that influences performance of process. Machining of deep, thin, and sharp-edged grooves is observed at an ideal rotation rate. It is critical to note that extremely high rotation speed reduces MRR (material removal rate) owing to challenges in film generation at such rates (Furutani & Maeda, 2008).

4.3.4 Electrochemical Discharge Dressing

It comprises of a worn grinding tool such as a cathode soaked in electrolyte and an auxiliary electrode called anode. Spark energy assists in the degrading of metal bonds and job debris bonds from the worn micro grinding tool surface, resulting in the protrusion of new grains around the micro grinding tool's perimeter. To avoid burs from machining regions, electrolyte is utilized as a dresser, cooling agent, dielectric medium, flushing agent, etc., in this technique

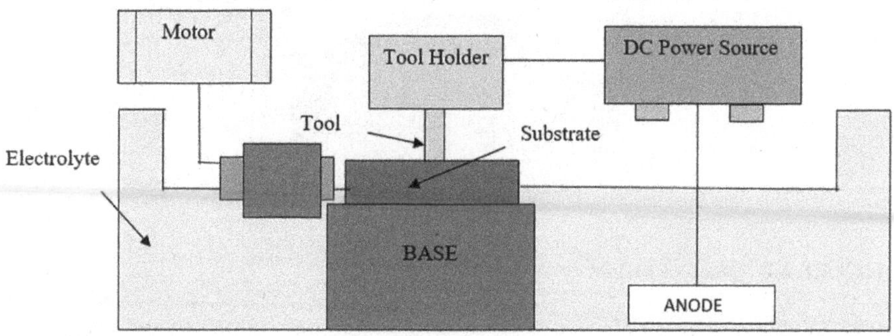

FIGURE 4.6 Electrochemical discharge turning.

(Wei & Hua, 2011). EDM is an unconventional metal removal technique for fulfilling the needs of today's production field by manufacturing geometries of complex shape and large variety of contemporary engineering materials. Several authors investigate the various properties of VAM (vibration-assisted machining processes), ultrasonic machining, rotary ultrasonic–based machining, etc., are used for the enhancement of machine surface by changing the input parameters, such as spindle speed, power, rate of feed, etc. (Sharma et al., 2019a, 2020, 2021b, 2022; Goud & Sharma, 2017).

4.3.5 Environmental Concern and Material Removal Mechanism

Electrochemical machining improves process accuracy, efficiency, automation, reliability, and environmental friendliness. Electrochemical etching of micro components for the computer-based sector has effectively used ECM ideas. EMM provides greater control and flexibility than traditional chemical etching, needs less monitoring, and has no serious safety or environmental problems (Datta, 1998; Datta & Romankiw, 1989). The need of ECDM and material removal mechanism are represented next in Figure 4.7. and Figure 4.8.

FIGURE 4.7 Need of ECDM.

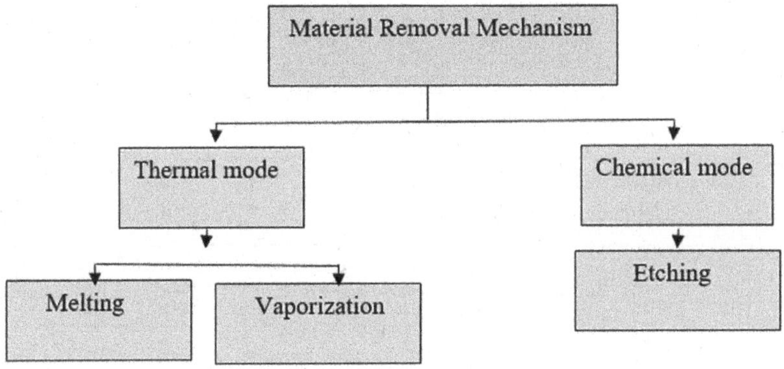

FIGURE 4.8 Material removal mechanism.

4.4 FUTURE SCOPE

The ECDM from the last ten years is progressively growing such as machining quality (tool travel rate, powder mixed electrolyte, etc.), hybridization (grinding-assisted ECDM, vibration-assisted ECDM, additive mixed ECDM, etc.), gas film (tool electrode shape, texture tool electrode, electrolyte temperature, etc.), tool electrode (flat side, needle shape, tubular shape etc.), and electrolyte (graphite-powder blend, silicon-powder blend, etc.). In addition, the distinct research scopes (Muniruddin & Ahmed, 2021; Basak & Ghosh, 1996) in ECDM are shown in Figure 4.9.

In thermal power plants, ECDM is used by several authors to machine turbine blades and other components made of nickel-based super alloys, which are difficult to machine by conventional methods due to their high hardness and toughness. ECDM offers several advantages over traditional machining methods, including high material removal rate, low tool wear, and excellent surface finish. In addition, in orthopedic implants, ECDM is used to machine titanium and titanium alloys, which are widely used due to their biocompatibility and excellent mechanical properties. ECDM offers several advantages over traditional machining methods, including high material removal rate, low thermal damage, and excellent surface finish. A study by A. J. Shetty et al. (2015) investigated the use of ECDM for machining of titanium alloys used in orthopedic implants. The study found that ECDM can achieve a high material removal rate of up to 2.4 mm3/min, with a surface roughness of 0.28 μm. The authors concluded that ECDM can be an effective alternative to conventional machining methods for the machining of titanium alloys in orthopedic implants. (Shetty et al., 2015; Kumar & Kumar, 2022; Kumar et al., 2021a, 2021b; Bedi et al., 2019).

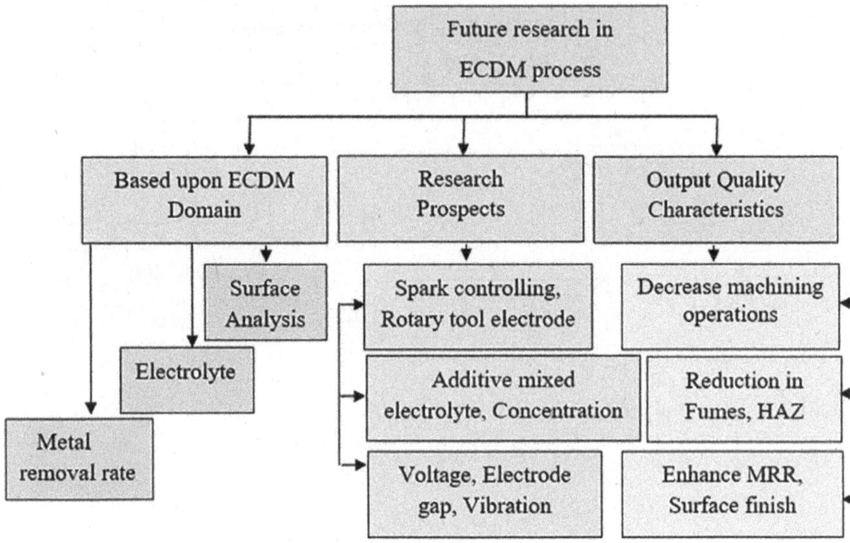

FIGURE 4.9 Future research scopes in ECDM process.

4.5 CONCLUSION

The ECDM termed as hybrid non-conducting metal removal process comprises the key features of EDM ECM. The removal of metal is by chemical dissolution as well as thermal melting. Thus, owing to the use of hard to brittle materials, this machining process plays a vital role in distinct sectors (aerospace, automotive, additive manufacturing electronics, etc.). Hence, a review has been performed to study the distinct materials and their future scope. However, wear is the major challenge to the current researchers or engineers. Hence, deep experimental study on material is required.

4.5.1 Declaration of Competing Interest

The authors declare that there is no financial aid for this paper.

4.5.2 Acknowledgement

The authors are extremely thankful to Chandigarh Group of Colleges' Landran, Mohali Punjab, and Auxein Medical Pvt. Ltd. Sonipat Haryana for offering the opportunity to carry out this research work.

REFERENCES

Abou Ziki, J.D., & Wüthrich, R. (2012). Tool wear and tool thermal expansion during micromachining by spark assisted chemical engraving. *Int. J. Adv. Manuf. Technol.*, 61, 481–486. https://doi.org/10.1007/s00170-011-3731-6

Akhai, S., & Rana, M. (2022). Taguchi-based grey relational analysis of abrasive water jet machining of Al-6061. *Mater. Today Proc.*, 65, 3165–3169. https://doi.org/10.1016/j.matpr.2022.05.361

Akhtar, W., Sun, J., & Sun, P. (2014). Tool wear mechanism in machining of Nickel based super-alloys: A review. *Front. Mech. Eng.*, 9(2), 106–119.

Ali, M.N., Doloi, B., & Sarkar, B.R. (2019). Electro-chemical discharge machining technology applied for turning operation. *IOP Conf. Ser. Mater. Sci. Eng.*, 653, 1–5.

Antil, P., Singh, S., & Manna, A. (2018a). Electrochemical discharge drilling of SiC reinforced polymer matrix composite using Taguch's Grey relational analysis. *Arab. J. Sci. Eng.*, 43(3), 1257–1266.

Antil, P., Singh, S., & Manna, A. (2018b). Glass fibers/SiCp reinforced epoxy composites: Effect of environmental conditions. *J. Compos. Mater.*, 52(9), 1253–1264.

Babbar, A., Jain, V., Gupta, D., Prakash, C., & Sharma, A. (2020a). Fabrication and machining methods of composites for aerospace applications. In *Characterization, Testing, Measurement, and Metrology* (1st ed., pp. 109–124). Boca Raton, FL: CRC Press.

Babbar, A., Jain, V., Gupta, D., Prakash, C., Singh, S., & Sharma, A. (2020b). Effect of process parameters on cutting forces and osteonecrosis for orthopedic bone drilling applications. In *Characterization, Testing, Measurement, and Metrology* (1st ed., pp. 93–108). Boca Raton, FL: CRC Press.

Babbar, A., Jain, V., Gupta, D., Prakash, C., Singh, S., & Sharma, A. (2020d). 3D bioprinting in pharmaceuticals, medicine, and tissue engineering applications. In *Advanced Manufacturing and Processing Technology* (1st ed., pp. 147–161). Boca Raton, FL: CRC Press, 2021.

Babbar, A., Jain, V., Gupta, D., & Sharma, A. (2020c). Fabrication of microchannels using conventional and hybrid machining processes. In *Non-Conventional Hybrid Machining Processes* (1st ed., pp. 37–51). Boca Raton, FL: CRC Press.

Babbar, A., Rai, A., & Sharma, A. (2021). Latest trend in building construction: Three-dimensional printing. *J. Phys. Conf. Ser.*, 1950, 012007.

Babbar, A., Sharma, A., & Singh, P. (2022). Multi-objective optimization of magnetic abrasive finishing using grey relational analysis. *Mater. Today Proc.*, 50, 570–575.

Basak, I., & Ghosh, A. (1996). Mechanism of spark generation during ECDM, a theoretical model and experimental verification. *J. Mater. Process. Technol.*, 62(1–3), 46–53. https://doi.org/10.1016/0924-0136(95)02202-3

Bedi, T.S., Kumar, S., & Kumar, R. (2019). Corrosion performance of hydroxyapaite and hydroxyapaite/titania bond coating for biomedical applications. *Mater. Res. Express*, 7(1), 015402. https://doi.org/10.1088/2053-1591/ab5cc5

Behroozfar, A., & Razfar, M.R. (2016). Experimental study of the tool wear during the electrochemical discharge machining. *Mater. Manuf. Process.*, 31, 574–580.

Bhattacharya, B., Doloi, B.N., & Sorkhel, S.K. (1999). Experimental investigation into electrochemical discharge machining (ECDM) of non-conducting materials. *J. Mater. Process. Tech.*, 95, 145–154.

Bhattacharyya, B., Munda, J., & Malapati, M. (2004). Advancement in electrochemical micro-machining. *Int. J. Mach. Tools Manuf.*, 44(15), 1577–1589.

Chak, S.K., & Rao, P.V. (2008). The drilling of Al2O3 using pulsed DC supply with a rotary abrasive electrode by the electrochemical discharge process. *Int. J. Adv. Manuf. Tech.*, 39, 633–641.

Charak, A., & Jawalka, C.S. (2019). A Theoretical analysis on electro chemical discharge machining using Taguchi method. *IOP Conf. Ser. J. Phys. Conf. Ser.*, 1240, 012083. https://doi.org/10.1088/1742-6596/1240/1/012083

Chavoshi, S.Z., & Behagh, A.M. (2014). A note on influential control parameters for drilling of hard-to-machine steel by electrochemical discharge machining. *Int. J. Adv. Manuf. Tech.*, 71, 1883–1887.

Coteata, M., Ciofu, C., & Slatineanu, L. (2009). Establishing the electrical discharge weight in electrochemical discharge drilling. *Int. J. Mater. Form.*, 2(1), 673–676.

Coteata, M., Schulze, H.P., & Slatineanu, L. (2011). Drilling of difficult-to-cut steel by electrochemical discharge machining. *Mater. Manuf. Process*, 26, 1466–1472.

Coteaţă, M., Slatineanu, L., & Dodun, O. (2008a). Electrochemical discharge machining of small diameter holes. *Int. J. Mater. Form.*, 1(1), 1327–1330.

Datta, M. (1998). Microfabrication by electrochemical metal removal. *IBM J. Res. Dev.*, 425, 655–669.

Datta, M., & Romankiw, L.T. (1989). Application of chemical and electrochemical micromachining in the electronics industry. *J. Electrochem. Soc.*, 136(6), 285c–292c.

Dhanvijay, M.R., & Ahuja, B.B. (2014). Micromachining of ceramics by electrochemical discharge process considering stagnant and electrolyte flow method. *Procedia Eng.*, 14, 165–172.

Furutani, K., & Maeda, H. (2008). Machining a glass rod with a lathe-type electro-chemical discharge machine. *J. Micromech. Micro Eng.*, 18, 8.

Goud, M., & Sharma, A.K. (2017). On performance studies during micromachining of quartz glass using electrochemical discharge machining. *J. Mech. Sci. Technol.*, 31(3), 1365–1372.

Gupta, P.K., Dvivedi, A., & Kumar, P. (2014). A study on the phenomenon of hole overcut with working gap in ECDM. *J. Prod. Eng.*, 17, 30–34.

Hofy, H.E., & Mcgeough, J.A. (1998). Evaluation of an apparatus for electrochemical arc wire-machining. *J. Eng. Ind.*, 110, 119–123.

Huang, S.F., Liu, Y., & Li, J. (2014). Electrochemical discharge machining micro-hole in stainless steel with tool electrode high-speed rotating. *Mater. Manuf. Process*, 29, 634–637.

Jain, V.K., & Adhikary, S. (2008). On the mechanism of material removal in electrochemical spark machining of quartz under different polarity conditions. *J. Mater. Process. Technol.*, 200(1–3), 460–470.

Jain, V.K., & Priyadarshini, D. (2014). Fabrication of micro-channels in ceramics (quartz) using ECSM. *J. Adv. Manuf. Syst.*, 13(1), 5–16.

Kalia, G., Sharma, A., & Babbar, A. (2022). Use of three-dimensional printing techniques for developing biodegradable applications: A review investigation. *Mater. Today Proc.* 62, 346–352.

Krotz, H., & Wegener, K. (2015). Spark assisted electrochemical machining: A novel possibility for micro drilling into electrical conductive materials using the electrochemical discharge phenomenon. *Int. J. Adv. Manuf. Tech.*, 79, 1633–1643.

Kumar, M., Kant, S., & Kumar, S. (2019). Corrosion behavior of wire arc sprayed Ni-based coatings in extreme environment. *Mater. Res. Express*, 6, 106427. https://doi.org/10.1088/2053-1591/ab3bd8

Kumar, R., Kumar, M., & Chohan, J.S. (2021a). Material-specific properties and applications of additive manufacturing techniques: A comprehensive review. *Bull. Mater. Sci.*, 44(3). https://doi.org/10.1007/s12034-021-02364-y

Kumar, R., Kumar, M., & Chohan, J.S. (2021b). The role of additive manufacturing for biomedical applications: A critical review. *J. Manuf. Process.*, 64, 828–850. https://doi.org/10.1016/j.jmapro.2021.02.022

Kumar, R., Kumar, M., & Singh, J. (2020a). The role of additive manufacturing for medical applications: A critical review. *J. Manuf. Process.*, 1–49

Kumar, R., & Singh, I. (2017). Electric discharge sawing of hybrid metal matrix composites. *Proc IMechE, Part B: Eng. Manuf.*, 231, 1775–1782.

Kumar, S., & Kumar, M. (2022). Tribological and mechanical performance of coatings on piston to avoid failure: A review. *J. Fail. Anal. Prev.* https://doi.org/10.1007/s11668-022-01436-3

Kumar, S., Kumar, M., & Handa, A. (2018). Combating hot corrosion of boiler tubes- a study. *J. Eng. Fail. Anal.*, 94, 379–395. https://doi.org/10.1016/j.engfailanal.2018.08.004

Kumar, S., Kumar, M., & Handa, A. (2020b). Erosion corrosion behaviour and mechanical property of wire arc sprayed Ni-Cr and Ni-Al coating on boiler steels in actual boiler environment. *Mater. High Temp.*, 37(6), 1–15. https://doi.org/10.1080/09603409.2020.1810922

Kumar, S., & Kumar, R. (2021). Influence of processing conditions on the properties of thermal sprayed coating: A review. *Surf. Eng.*, 37(11), 1339–1372. https://doi.org/10.1080/02670844.2021.1967024

Kunze, J.M., & Bampton, C.C. (2001). Challenges to developing and producing MMCs for space applications. *JOM*, 53(4), 22–25.

Kurafuji, H., & Suda, K. (1968). Electrical discharge drilling of glass. *Ann CIRP*, 16, 415–419. https://doi.org/10.2493/jjspe1933.26.596

Liao, Y.S., & Peng, W.Y. (2006). Study of hole-machining on pyrex wafer by electrochemical discharge machining (ECDM). *Mater. Sci. For.*, 505, 1207–1212.

Liu, J.W., Yue, T.M., & Guo, Z.N. (2010). An analysis of the discharge mechanism in electrochemical discharge machining of particulate reinforced metal matrix composites. *Int. J. Mach. Tool. Manu.*, 50, 86–96.

Lu, Y., Liu, K., & Zhao, D. (2011). Experimental investigation on monitoring interelectrode gap of ECM with six-axis force sensor. *Int. J. Adv. Manuf. Technol.*, 55, 565–572. https://doi.org/10.1007/s00170-010-3105-5

Malik, A., & Manna, A. (2016). An experimental investigation on developed WECSM during micro slicing of e-glass fiber epoxy composite. *Int. J. Adv. Manuf. Tech.*, 85, 2097–2106.

Manna, A., & Kundal, A. (2015). An experimental investigation on fabricated TW-ECSM setup during micro slicing of nonconductive ceramic. *Int. J. Adv. Manuf. Tech.*, 76, 29–37.

Manna, A., & Narang, V. (2012). A study on micro machining of e-glass fibre—epoxy composite by ECSM process. *Int. J. Adv. Manuf. Technol.*, 61, 1191–1197.

Min-Seop, H., & Byung-Kwon, M. (2009). Geometric improvement of electrochemical discharge micro-drilling using an ultrasonic-vibrated electrolyte. *J. Micromech. Microeng.*, 19, 065004.

Muniruddin, M.M.S., & Ahmed, M.I.M. (2021). Review paper on electrochemical discharge machining: Mechanism and future scope. *Int. J. Eng. Res. Technol. (IJERT)*, 20(4), 104–106. https://doi.org/10.17577/IJERTCONV9IS04024

Neto, S., & Cirilo, J. (2011). Development of a prototype of electrochemical machining. In *17th CIRP Conference on Modelling of Machining Operations*. Sintra, Portugal: Trans Tech Publications.

Ozkeskin, F.M. (2008). Feedback controlled high frequency electrochemical micromachining. In *54th International Instrumentation Symposium*. Pensacola Beach, FL: ISA—Instrumentation, Systems, and Automation Society.

Parikh, P., Sharma, A., Trivedi, R., Roy, D., & Joshi, K. (2023). Performance evaluation of an indigenously-designed high-performance dynamic feeding robotic structure using advanced additive manufacturing technology, machine learning and robot kinematics. *Int. J. Interact. Des. Manuf. (IJIDeM)*, 1–29.

Paul, L., & Hiremath, S.S. (2013). Response surface modelling of micro holes in electrochemical discharge machining process. *Procedia Eng.*, 64, 1395–1404.

Prakash, C., Kumar, V., Mistri, A., Sharma, A., Uppal, A.S., Babbar, A., & Pathri, B.P. (2021). Investigation of functionally graded adherents on failure of socket joint of FRP composite tubes. *Materials*, 14, 6365.

Rampal, R., Goyal, T., Goyal, D., Mittal, M., Dang, R.K., & Bahl, S. (2021). Magneto-rheological abrasive finishing (MAF) of soft material using abrasives. *Mater. Today Proc.*, 45, 51140–5121. https://doi.org/10.1016/j.matpr.2021.01.629

Rana, M., & Akhai, S. (2022). Multi-objective optimization of Abrasive water jet Machining parameters for Inconel 625 alloy using TGRA. *Mater. Today Proc.*, 65, 3205–3210. https://doi.org/10.1016/j.matpr.2022.05.374

Sarkar, B.R., Doloi, B., & Bhattacharyya, B. (2006). Parametric analysis on electrochemical discharge machining of silicon nitride ceramics. *Int. J. Adv. Manuf. Technol.*, 28, 873–881. https://doi.org/10.1007/s00170-004-2448-1

Sharma, A., Babbar, A., Jain, V., & Gupta, D. (2018a). Enhancement of surface roughness for brittle material during rotary ultrasonic machining. *MATEC Web Conf.*, 249, 01006.

Sharma, A., Fidan, I., Huseynov, O., Ali, M.A., Alkunte, S., Rajeshirke, M., Gupta, A., Hasanov, S., Tantawi, K., Yasa, E., Yilmaz, O., Loy, J., & Popov, V. (2023a). Recent inventions in additive manufacturing: Holistic review. *Inventions*, 8(4), 103. https://doi.org/10.3390/inventions8040103

Sharma, A., Grover, V., Babbar, A., & Rani, R. (2020b). A trending nonconventional hybrid finishing/machining process. In *Non-Conventional Hybrid Machining Processes* (1st ed., pp. 79–93). Boca Raton, FL: CRC Press.

Sharma, A., & Jain, V. (2020). Experimental investigation of cutting temperature during drilling of float glass specimen. In *IOP Conference Series: Materials Science and Engineering* (Vol. 715, No. 1, p. 012050). IOP Publishing. https://doi.org/10.1088/1757-899X/715/1/012050

Sharma, A., Jain, V., & Gupta, D. (2018b). Characterization of chipping and tool wear during drilling of float glass using rotary ultrasonic machining. *Measurement*, 128, 254–263.

Sharma, A., Jain, V., & Gupta, D. (2019a). Comparative analysis of chipping mechanics of float glass during rotary ultrasonic drilling and conventional drilling: For multi-shaped tools. *Mach. Sci. Technol.*, 23, 547–568.

Sharma, A., Jain, V., & Gupta, D. (2019b). Multi-shaped tool wear study during rotary ultrasonic drilling and conventional drilling for amorphous solid. *Proc. Inst. Mech. Eng. Part E J. Process. Mech. Eng.*, 233, 551–560.

Sharma, A., Jain, V., & Gupta, D. (2019c). Tool wear analysis while creating blind holes on float glass using conventional drilling: A multi-shaped tools study. In *Advances in Manufacturing Processes: Select Proceedings of ICEMMM 2018* (pp. 175–183). Singapore: Springer. https://doi.org/10.1007/978-981-13-1724-8_17

Sharma, A., Jain, V., & Gupta, D. (2022). Mathematical approach on chipping volume estimation generated during rotary ultrasonic drilling for float glass. *Proc. Natl. Acad. Sci. India Sect. A—Phys. Sci.*, 92, 285–291.

Sharma, A., Jain, V., Gupta, D., & Babbar, A. (2020a). A review study on miniaturization. In *Advanced Manufacturing and Processing Technology* (1st ed., pp. 111–131). Boca Raton, FL: CRC Press.

Sharma, A., Kalsia, M., Uppal, A.S., Babbar, A., & Dhawan, V. (2021b). Machining of hard and brittle materials: A comprehensive review. *Mater. Today Proc.* https://doi.org/10.1016/j.matpr.2021.07.452

Sharma, A., Kumar, V., Babbar, A., Dhawan, V., Kotecha, K., & Prakash, C. (2021c). Experimental investigation and optimization of electric discharge machining process parameters using grey-fuzzy-based hybrid techniques. *Materials*, 14, 5820. https://doi.org/10.3390/ma14195820

Sharma, A., Sandhu, H.S., Goyal, D., Goyal, T., Jarial, S., & Sharda, A. (2023b). Sustainable development in cold gas dynamic spray coating process for biomedical applications: Challenges and future perspective review. *Int. J. Interact. Des. Manuf. (IJIDeM)*, 1–17. https://doi.org/10.1007/s12008-023-01474-7

Sharma, V.K., Rana, M., Singh, T., Singh, A.K., & Chattopadhyay, K. (2021d). Multi-response optimization of process parameters using Desirability Function Analysis during machining of EN31 steel under different machining environments. *Mater. Today Proc.*, 44, 3121–3126. https://doi.org/10.1016/j.matpr.2021.02.809

Shetty, A.J., Kumar, P., & Raju, B. (2015). Experimental investigation of electrochemical discharge machining for biomedical applications. *Procedia CIRP*, 31, 65–70.

Silva, A.D., & McGeough, J.A. (1986). Surface effects on alloys drilled by electrochemical arc machining. *Proc. Inst. Mech. Eng. Part B J. Eng. Manuf.*, 237–246.

Singh, G., Kumar, S., & Kumar, R. (2020). Comparative study of hot corrosion behaviour of thermal sprayed alumina and titanium oxide reinforced alumina coatings on boiler steel. *Mater. Res. Express*, 7, 026527. https://doi.org/10.1088/2053-1591/ab6e7e

Singh, R.P., & Singhal, S. (2016). An experimental study on rotary ultrasonic machining of Macor ceramic. *Proc. IMechE., Part B J. Eng. Manuf.*, 232(7), 1221–1234.

Singh, S., Singh, I., & Dvivedi, A. (2013a). Design and development of abrasive-assisted drilling process for improvement in surface finish during drilling of metal matrix composites. *Proc. IMechE., Part B J. Eng. Manuf.*, 228(8), 857–867.

Singh, S., Singh, I., & Dvivedi, A. (2013b). Multi objective optimization in drilling of Al6063/10% SiC metal matrix composite based on grey relational analysis. *Proc. IMechE, Part B J. Eng. Manuf.*, 227(12), 1767–1776.

Singh, T., & Dvivedi, A. (2016). Developments in electrochemical discharge machining: A review on electrochemical discharge machining, process variants and their hybrid methods. *Int. J. Mach. Tools Manuf.*, 105. http://doi.org/10.1016/j.ijmachtools.2016.03.004

Singh, Y.P., Jain, V.K., & Kumar, P. (1996). Machining piezoelectric (PZT) ceramics using an electrochemical spark machining (ECSM) process. *J. Mater. Process Tech.*, 58, 24–31.

Speer, W., & Es-Said, O.S. (2001). Applications of an aluminium-beryllium composite for structural composite for structural aerospace components. *Eng. Fail. Anal.*, 2004, 11, 895–902.

Tandon, S., Jain, V.K., & Kumar, P. (1990). Investigation into machining of composites. *Precis. Eng.*, 12, 227–238.

Taweeporn, W., Viboon, T., & Chaiya, D. (2015). Lascr micromilling under a thin and flowing water layer: A new concept of liquid-assisted laser machining process. *Proc. IMechE. Part B J. Eng. Manuf.*, 230(2), 376–380.

Wang, X., Zhao, D., & Yun, N. (2007). Research on intelligent measurement and control method of interelectrode gap of electrochemical machining (ECM). *China Mech. Eng.*, 18(23), 2860–2864.

Wei, C., & Hua, D. (2011). Electro chemical discharge dressing of metal bond icrogrinding tools. *Int. J. Mach. Tools Manuf.*, 51, 165–168.

West, J., & Jadhav, A. (2007). ECDM methods for fluidic interfacing through thin glass substrates and the formation of spherical micro cavities. *J. Micromech. Micro Eng.*, 17, 403–409.

Wuthrich, R., & Fascio, V. (2005). Machining of non-conducting materials using electrochemical discharge phenomenon—an overview. *Int. J. Mach. Tools Manuf.*, 45(9), 1095–1108. https://doi.org/10.1016/j.ijmachtools.2004.11.011

Yadav, R.N., & Yadava, V. (2017a). Experimental investigations of slotted electrical discharge abrasive grinding of Al/SiC/Gr composite. *Proc IMechE Part B J Eng. Manuf.*, 231, 945–955.

Yadav, U.S., & Yadava, V. (2017b). Experimental investigation on electrical discharge diamond drilling of nickel-based superalloy aerospace material. *Proc. IMechE Part B J. Eng. Manuf.*, 231(7), 1160–1168.

Yan, Z., Zhengyang, X., & Jun, X. (2016a). Effect of tube electrode inner diameter on electrochemical discharge machining of nickel-based superalloy. *Chinese J. Aeronaut.*, 29(4), 1103–1110.

Yan, Z., Zhengyang, X., & Yun, Z. (2016b). Machining of film cooling hole in a single-crystal superalloy by high-speed electrochemical discharge machining. *Chinese J. Aeronaut.*, 29(2), 560–570.

Yang, C.T., Song, S.L., Yan, B.H., & Huang, F.Y. (2006). Improving machining performance of wire electrochemical discharge machining by adding SiC abrasive to electrolyte. *Int. J. Mach. Tools Manuf.*, 46, 2044–2050. https://doi.org/10.1016/j.ijmachtools.2006.01.006

Yuan, S., Zhu, G., & Zhang, C. (2017). Modeling of tool blockage condition in cutting tool design for rotary ultrasonic machining of composites. *Int. J. Adv. Manuf. Tech.*, 91(5–8), 2645–2654.

Zhang, Y., Xu, Z., & Zhu, Y. (2016). Effect of tube-electrode inner structure on machining performance in tube electrode high-speed electrochemical discharge drilling. *J. Mater. Process Tech.*, 231, 38–49.

Zheng, Z.P., & Cheng, W.H. (2007). 3D micro structuring of Pyrex glass using the electrochemical discharge machining process. *J. Micromech. Micro Eng.*, 960–966.

Ziki, J.D.A., & Wüthrich, R. (2013). Spark assisted chemical engraving. *Int. J. Mach. Tools Manuf.*, 73, 47–54.

Zweben, C. (2005). Advanced electronics packaging material. *Adv. Mater. Process.*, 163, 33–37.

5 Advancement of Abrasive-Based Nano-Finishing Processes

Principle, Challenges, and Current Applications

Manoj Kumar, Mohit Kumar, Ankit Sharma, Atul Babbar

Nomenclature and Notations

AFM	abrasive flow machining
AJM	abrasive jet machining
AWJM	abrasive water jet machining
BEMRF	ball end magnetorheological finishing
CMP	chemical mechanical polishing
EBM	electron beam machining
ECM	electrochemical machining
ECM	electrochemical machining
EDM	electric discharge machining
EDM	electric discharge machining
LBM	laser beam machining
MEMS	micro electromechanical systems
MMC	metal matrix composite
MRAFF	magnetorheological fluid abrasive flow finishing process
MRAH	magnetorheological abrasive honing
MRF	magnetorheological finishing
MR	magnetorheological
MRR	material removal rate
PSD	power spectral density
SR	surface roughness
SSD	sub-surface damage
WC	tungsten carbide

DOI: 10.1201/9781003327905-5

5.1 INTRODUCTION

The surface quality mainly depends on the surface properties achieved by surface finishing operations. This has drawn the researchers' attention toward the surface integrity of the machined or processed part. The surface quality greatly influences product applicability in different domains (Loveless et al., 1994; Shi and Gibson, 1998; Hassanin et al., 2018; Zou et al., 2020; Prakash et al., 2021; Babbar et al., 2022a; Sharma et al., 2022). The researchers constantly work toward the nano-processing and finishing the other materials with certain accuracy and precision and accuracy for meeting the functional requirement (Jefferies, 2007; Heng et al., 2022). Depending on the geometry, the surface might be of different orientations, such as plain surface, curved surface, or freeform surface (Fang et al., 2013). These differences surface a different set of challenges for researchers in this area. The freeform surface has no rotational axis and is difficult to finish at the nanoscale compared to the plain or other symmetrical parts or components. In recent times, the importance of surface characterization has improved a lot. As per the research, the finishing operations cost around 15% of the total cost of the product. The surface quality is mainly determined through surface integrity which includes several controllable parameters (Jain, 2008). Further, few studies were reported while machining ductile and brittle materials using various machining techniques which shows a futuristic approach to machining techniques (Babbar et al., 2022, 2021a, 2021b, 2021c, 2020; Sharma et al., 2022a, 2022b, 2022c, 2021b, 2020, 2019a, 2019b, 2019c, 2018; Sharma & Jain, 2020b; Singh et al., 2021, 2022).

As discussed, the finishing operations are the secondary operations carried out by either initial subtractive processes (machining) or additive processes (3D-printed components) (Lee et al., 2021b; Ye et al., 2021; Annamaria et al., 2022). These are the costlier operations compared to the primary manufacturing methods, such as casting or forming. Finishing operations require high skill and experience to operate highly sophisticated machines and metrological instruments (Debnath et al., 2017). The workpiece must process at an excellent level with certain accuracy and precision. Most processes have mechanical-based material removal mechanisms, such as abrasive flow machining. The mechanical force generated on the abrasive particle removes the material during this process. The abrasive particle is responsible for material removal. The process is best suited for finishing the different geometries. However, the complex geometry's fragile part may fail under mechanical force. Therefore, the mechanical-based material removal processes have limited use in such cases (Kumar et al., 2022a). However, chemical etching can remove the material under various chemical phenomena. Faraday's laws govern material removal in the process. The chemical reactions govern the material removal. The material removal method can be further improved using photochemical-based methods (Allen, 2003). In this method, the substrate surface is prepared along with the photo tool. The materials can be finished by making functional changes in the conventional grinding by deploying electrochemical phenomena (Wangikar et al., 2019; Babbar et al., 2020). The negatively charged abrasive particles remove the fine material from the positive workpiece in the presence of a chemical-based electrolyte. Further, highly sophisticated processes like laser beam and electron beam machining can be used to finish the material. In this process, the material is removed by melting the workpiece and vaporing through high temperatures—the high power density and cost involved in these processes (Dubey and Yadava, 2008; Sahu et al., 2022). The different types of material removal mechanism have been summarized in the Figure 5.1.

The surface roughness is one of the significant response parameters to characterize the surface finish. Broadly, the surface roughness is characterized by 2D surface roughness, i.e. (Ra) and 3D surface (Sa) (Singh et al., 2017b). The 2D surface roughness is measured through a contact-type stylus probe, and the 3D surface is measured through a no-contact type 3D profilometer. Further, the finished surface is also characterized by waviness, power spectral density, etc. (Kumar et al., 2021b). The surface burr, residual or surface stress caused due to initial or primary manufacturing processes, also affects the surface finish (Singh et al., 2017a; Khatri et al., 2018).

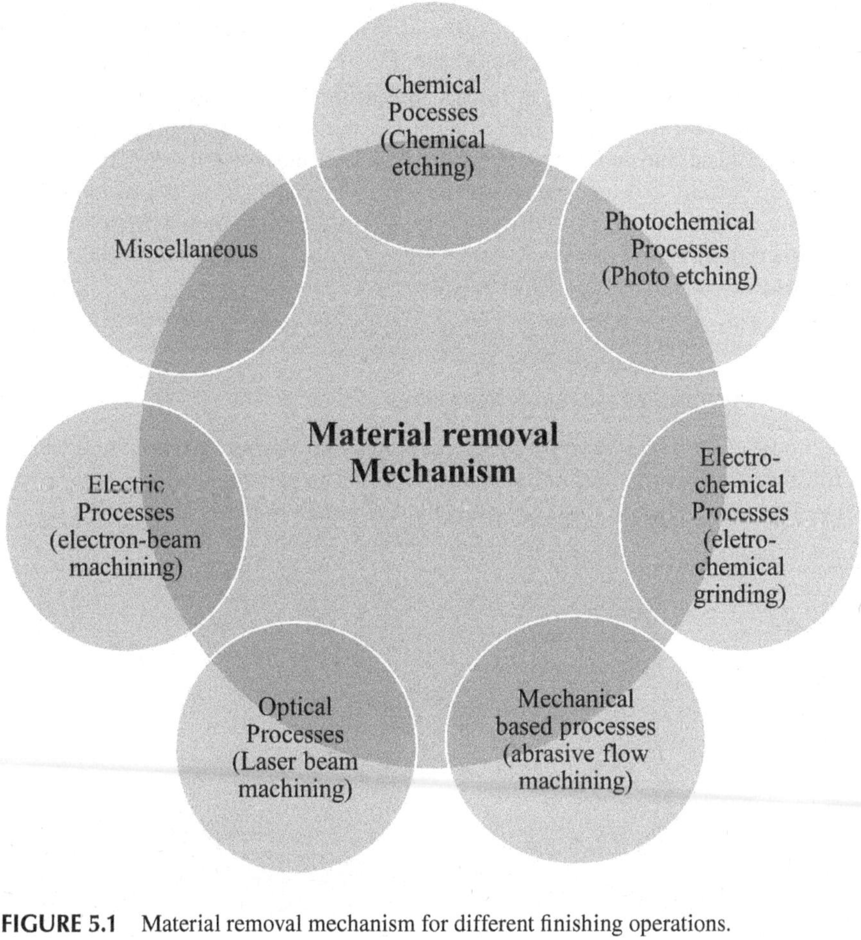

FIGURE 5.1 Material removal mechanism for different finishing operations.

5.2 DIFFERENT ABRASIVE-BASED NANO-FINISHING PROCESSES

Any product's shape, size, dimensional tolerances, surface integrity, and surface finish are crucial characteristics. The final process on the part, referred to as the finishing operation, produces the surface finish. It also determines other aspects of the machined part's surface in most situations, such as surface flaws like micro cracks (Kumar et al., 2020b; Kumar and Satsangi, 2021). Finishing operations that increase workpiece temperature above permitted may cause warping, residual thermal stresses, and surface flaws like microscopic cracks. Therefore, choosing the appropriate finishing procedure is crucial. Understanding the many different finishing operations that are accessible is also essential for this purpose. In light of this, this material removal finishing processes like traditional finishing processes (like honing, lapping, superfinishing, burnishing, polishing, and buffing) and advanced finishing processes such as magnetic abrasive finishing, magnetic float polishing, elastic emission machining, and ion beam machining) and material addition processes like electroplating, galvanizing, and metal spraying (Kumar et al., 2021a; Sharma et al., 2021; Kumar and Singh, 2022).

There are different ways to categorize the finishing processes. However, the finishing operations are divided into two parts: traditional and advanced machining (Figure 5.2). The conventional machining methods include grinding, lapping, honing, polishing, buffing, and superfinishing. In this process, a mechanical force dominates under the action of an abrasive that might be independent or bonded (Mizobuchi and Tashima, 2020). These processes have different surface finishes under other conditions. However, buffing and superfinishing processes are considered to achieve a surface finish of up to 0.01 μm.

5.2.1 Traditional Finishing Operations

The high-speed spinning polishing wheel is made of leather or fabric; polishing is a finishing process to improve the surface finish. The polishing wheel's outer rim is coated with an adhesive that holds the abrasive grains in place. Polishing tasks are

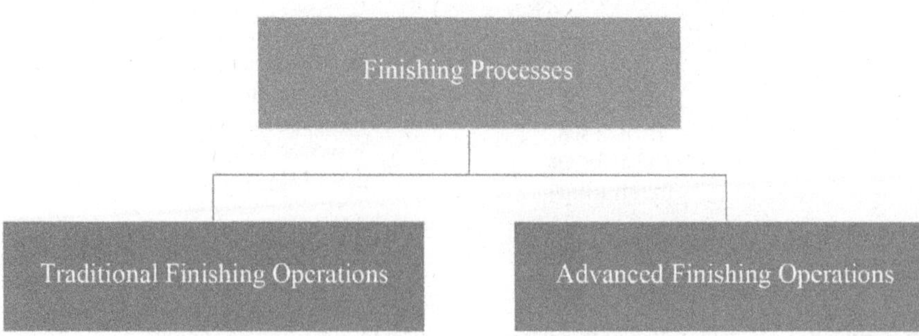

FIGURE 5.2 Classifications of finishing operations.

frequently carried out manually. Like polishing, buffing is a finishing process where abrasive granules are enclosed in a buffing compound that is forced into the outside surface of the buffing wheel as it revolves. The abrasive particles need to be replaced regularly, much like while polishing. Like polishing, buffing is typically carried out by hand, though specific machines can perform the task automatically. When a better surface polish is required than what can be attained through grinding and honing, lapping is used. High dimensional precision, shape correction of minor flaws, and a tight fit between mating surfaces are all achieved through its use. However, the operation is more expensive than honing and grinding. Lapping involves switching between a lap (often formed of soft material) and the workpiece loose suspended abrasives (Zhong, 2020).

Except that buffing uses excellent abrasives on soft discs made of cloth, it is comparable to polishing. The buffing wheel is coated with the abrasive grains in a suitable carrying medium, such as grease, at appropriate intervals. Buffing can get an even finer surface quality on polished components. Although little material is lost during buffing, the finished surface has a high degree of shine. The buffed parts' dimensional accuracy is unaffected. It should be free of flaws and severe scratches for a mirror finish. The abrasive mixed-in lubricating oil/grease is usually aluminium oxide. By forcing elevated micro-irregularities or peaks on the surface into tiny crevices, the burnishing of metals is a technique for creating smooth surfaces. The surface that was created has a high level of finish. Burnishing is done with a ball- or roller-type instrument that moves when applied pressure to a surface-revolving piece of work (Kalisz et al., 2021).

The metal is plastically deformed and pressed into the valleys by the tool as it progresses. Finally, a completed surface that is smooth is achieved. Due to strain hardening caused by the surface deformation of the micro-irregularities, burnishing also increases the workpiece's surface hardness (Vaishya et al., 2022).

5.2.2 Advanced Finishing Operations

Numerous developments are being made in the fine abrasive finishing of materials, including procedures, abrasives, and their bonding, enabling them to provide surface finishes on the order of nanometers. Abrasives could only be made as delicate as a few micrometers before, but modern breakthroughs in material synthesis have made it possible to produce materials as small as a nanometer. For the purpose of manufacturing components with various shapes, sizes, accuracy, finishes, and surface integrity, abrasives are employed in various forms, including loose abrasives (polishing, lapping), bonded abrasives (grinding wheels), and coated abrasives. The detailed advanced finishing processes have been shown in the Figure 5.3.

Recent advancements in advanced finishing techniques, such as EDM, ECM, ECDM USM, and AJM, have relaxed the requirements for tool hardness (Kumar et al., 2020a, 2022b). The cutting edge's predetermined relative motion to the workpiece surface is a significant barrier to completing complicated designs. Weakly bonded materials include several cutting edges directed to finish the complex shapes. However, due to the finishing forces' lack of control, these processes are limited in their ability to complete complex geometries, and occasionally, they cause surface and sub-surface damage. Such issues have been addressed through numerous cutting-edge finishing techniques.

5.3 ABRASIVE FLOW MACHINING (AFM) PROCESS

The abrasive flow machining (AFM) process is an advanced finishing technique that removes the minor surface defects developed in the primary manufacturing technique. The process can remove surface irregularities such as residual stress, burr, craters, waviness, and geometric misalignment. Intense research has been done in this domain for the last couple of decades. Various hybridizations have been adopted to change and improve the functionality of the process (Dixit et al., 2021a, 2021b).

In the AFM process, a semi-solid media is used that consist of a carrier in the form of a polymer base that contains abrasive particles in the required proportion and is extruded across the surface that needs to be machined under the specified

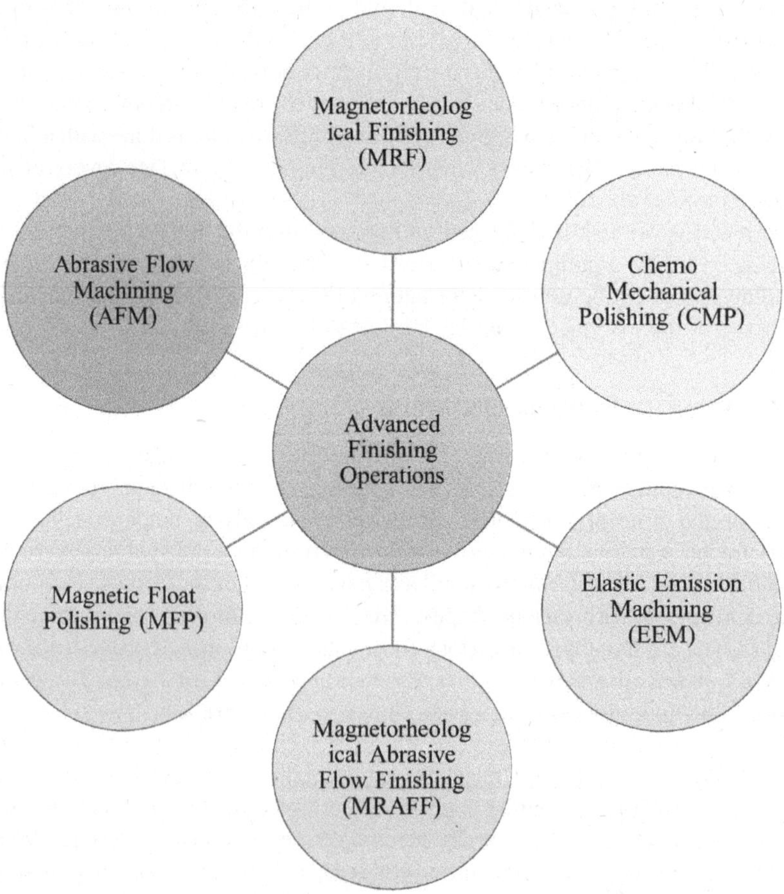

FIGURE 5.3 Classifications of advanced finishing operations.

pressure. There are different types of mechanism involves in the abrasive flow finishing process (Figure 5.4). When the media is subjected to limitations because of an uneven surface, it functions as a flexible instrument. Media has a unique malleable capability that allows it to move through any shape of passage.

At the surfaces that AFM will process, restricted media flow pathways are required. The media deports remotely like a flexible grinding stone, abrades the material, and produces a good surface finish. Typically, a fixture is needed to restrict or focus and direct the media to a specific zone of the workpiece.

Three categories of abrasive flow finishing (AFF) setup configurations have been established: (1) one-way AFF, (2) two-way AFF, and (3) orbital AFF (Figure 5.5) (Dixit et al., 2021a). In a one-way AFF process, an AFM media cylinder and a hydraulically driven reciprocating piston cylinder are positioned so that the AFF media flow unidirectional across the internal surfaces of the workpiece. In a two-way AFF arrangement, two vertically positioned media chambers extrude the AFF media over the workpiece surfaces in both directions. Within a slowly flowing pad of elastic or plastic AFF media, the workpiece exactly oscillates in two or three proportions when in orbital AFF. In addition to these traditional AFF configurations, the literature also mentions several hybridized AFF processes, such as elastic emission machining (EEM), drill bit-guided abrasive flow finishing (DBG-AFF), and centrifugal force-assisted. Further, there are lots of research potential that needs to exploits in detailed as summarized in the Figure 5.6.

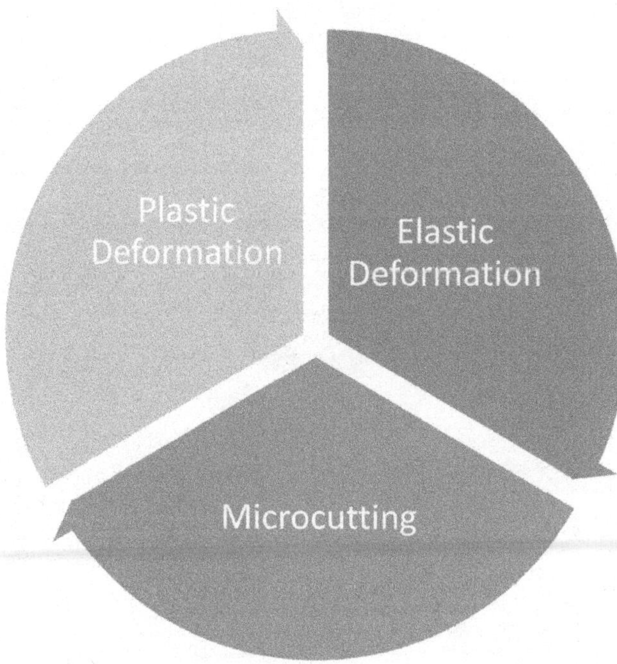

FIGURE 5.4 Modes of material removal mechanism.

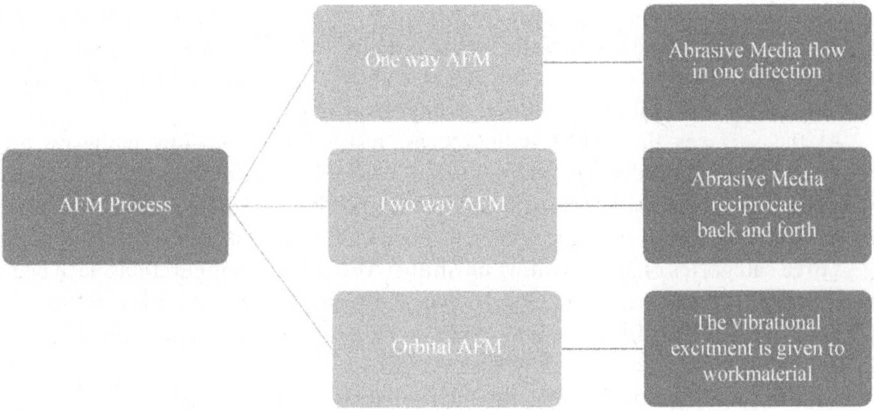

FIGURE 5.5 Broad classifications of the AFM process.

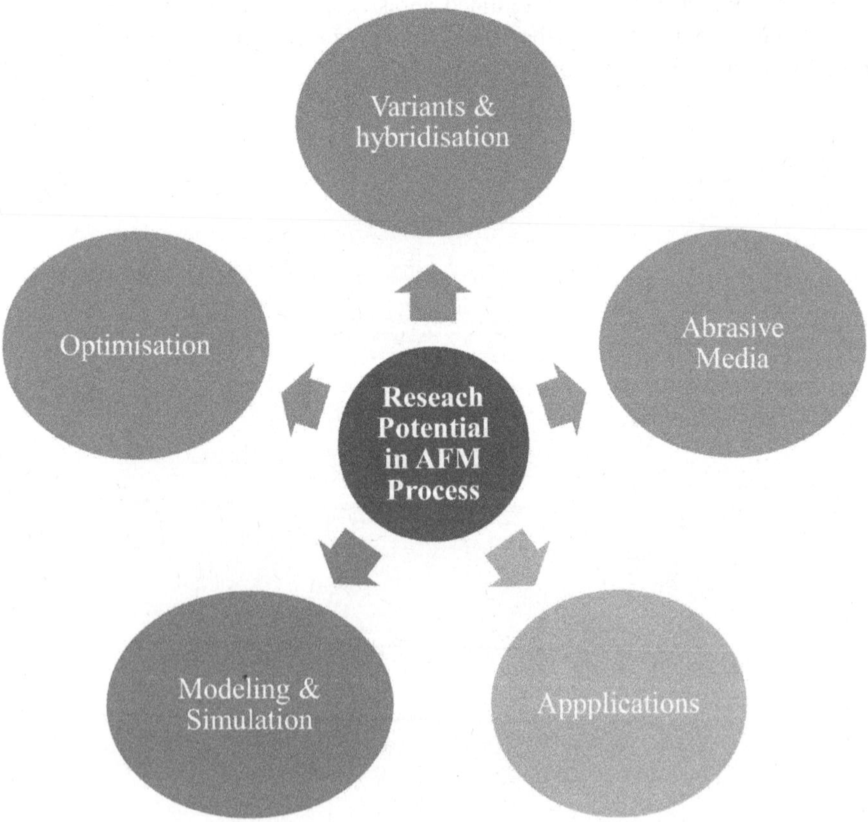

FIGURE 5.6 Research Potential in AFM process.

The details of the experimental investigations performed on the developed variants of AFM are given in Table 5.1.

5.3.1 Abrasive Media

The rheology of abrasive media plays a vital role during the finishing process. Therefore, the rheology of abrasive media needs to be characterized before its use. Most of the abrasive media have shear thinning behavior. The rheology of abrasive media is affected by the type of ingredients (polymer base, processing oil, and abrasives), the concentration of ingredients, temperature, and shear rate. The viscosity of abrasive media decreases with an increase in abrasive mesh size, shear rate, wall shear stress, and temperature rise during machining (Jain et al., 2001). Higher abrasive concentration also found to result in a higher viscosity of media (Singh et al., 2019). In this regard, Kar et al. (2009a) reported that BR has a more rigid structure than NR. Therefore, NR based media produced better finishing results and can bear more stress for given strain value. In an another research attempt, Fang et al. (2009) performed the experiments on AISI 1080, 1045, and A36 steels. It was observed that media with higher viscosity resulted in better Ra and MR than low viscosity media. Temperature rise during machining affects the finishing performance adversely. In experimental studies, it was reported that in AFM, temperature usually rises to 45–50°C (Sankar et al., 2011). This temperature rise resulted in a decrease in media viscosity. Therefore, thermal analysis of abrasive media is also very much important. The thermal analysis performed for different rubber-based media shows good thermal stability of abrasive media, as indicated by TGA (Kar et al., 2009b). The SBR based abrasive media was reported to have maximum degradation temperature, even greater than commercial abrasive media. The natural polymer-based abrasive media was also reported to be stable till a temperature of 412 K (Gudipadu et al., 2015). The useful life of abrasive media is affected by initial surface quality of work material, abrasive mesh size, flow rate, and part geometry [12]. The wear of media during the machining process was investigated by Bremerstein et al. (2015). It was found that the use of already used media for finishing resulted in 20% poor surface quality and 30% less material removal. The bluntness of abrasives is calculated in terms of change in cutting-edge radius. Singh et al. (2016) measured the bluntness of abrasive media with extrusion pressure and the number of cycles. The author reported that an increase in extrusion pressure resulted in an increase in more radial force as compared to the axial force component. This caused rolling of abrasive particles and reduction in their sharpness. In addition to media development and characterization, detailed literature about the performance analysis of different abrasive media used in AFM is given in Table 5.2.

5.4 CHEMICAL MECHANICAL POLISHING (CMP)

Chemical mechanical polishing (CMP) is an advanced finishing technique capable and used to finish the surface of almost all types of materials, irrespective of conductivity (Pal et al., 2022b). The process is extensively utilized for glass materials. Different glass materials, such as borosilicate glass, quartz glass, soda-lime glass, etc.,

TABLE 5.1
Experimental Investigations Performed on the Developed Variants of AFM

Reference	Variant	Work Material	Variant-Based Parameter	Variant Performance
			Ultrasonic Assistance	
Jones and Hull (1998)	UFP	Aluminium	• Amplitude of vibration (μm) = 4.5 • Frequency (kHz) = 40 • Abrasive = boron carbide	• UFP reduced the surface roughness by a factor of 10:1. • More homogeneous surface obtained in manual polishing compared to UFP.
Sharma et al. (2015)	UAAFM	EN8 steel	• Media viscosity (Pa-s) = 730 • Applied frequency (kHz) = 0, 5, 10, 15, 20 • Amplitude of vibration (μm) = 10	• Best surface finish was obtained at a frequency of 15 kHz. • The ultrasonic frequency was reported to be the most significant parameters for % ΔRa. • Maximum % ΔRa was 81.02% at 7 kgf/cm2 extrusion pressure. • Maximum MR was 14.5 mg after seven minutes of finishing. • Vibrations at the work surface also cause a change in microstructure due to straining at tap layer.
Venkatesh et al. (2015)	UAAFM	EN8 steel	• Frequency of vibration (kHz) = 19 • Amplitude of vibration (μm) = 10	• For the same finishing time % ΔRa was 55% for AFM and 73.12% in UAAFM. • % ΔRa and MR for five minutes of machining using UAAFM was found to be higher than 15 minutes of machining using AFM.
			Magnetic Assistance	
Singh et al. (2002)	MAFM	Brass, aluminium, mild steel	Magnetic flux density (MFD) (T) = 0 and 0.7	• In brass, material removal occurs due to abrasion, but there is smearing in Al with low abrasion. • No effect of the magnetic field was observed on MR and Ra for mild steel due to the shielding effect. • The magnetic field has 88.87% contribution in surface roughness for brass material and only 2.41% for aluminium.
Singh et al. (2020a)	MAFM	Aluminium	• MFD (T) = 0–0.75 • No. of cycles = 10–30 • Extrusion pressure (MPa) = 1.4–7	• Developed magnetic abrasives by mixing diamond powder and iron powder followed by mechanical alloying. • Highest % ΔRa was 72.7% at 0.6 T, 25 cycles, at 5.6 MPa extrusion pressure. • The best surface finish obtained was 0.22 μm.

Magnetorheological Variants				
Jha and Jain (2004)	MRAFF	Mild steel	• MFD (T) = 0, 0.152, 0.388, 0.531, 0.574	• No measurable change in Ra in absence of magnetic field. • Improvement in Ra increased with an increase in magnetic flux density. • Abrasive indentation marks were observed at high magnetic flux density due to deeper penetration.
Ghadikolaei and Vahdati (2015)	MRAFF	Copper, aluminium, SS316	• MFD (T) = 0.4, 0.8, and 1.2 • Cycle time (hr.) = 0.5, 1, and 1.5	• The surface quality decrease with increase in MFD due to increased penetration of abrasives under high magnetic field. • Surface roughness decreases only up to certain finishing time and then again start increasing due to scratches by abrasive particles.
Das et al. (2010)	RMRAFF	Stainless steel	• Rotational speed of the magnet (rpm) = 50, 100, 150, 200, 250	• Increase in the rotation speed of magnet results in shear thinning of media, thus reducing its performance. • The surface finish of 16 nm obtained after finishing. • The crosshatched pattern was obtained after finishing like the honing process.
Das et al. (2012a)	RMRAFF	EN8, stainless steel, brass	• Rotational speed of the magnet (rpm) = 20, 40, 60, 80, 100 • Volume ratio of CIP/SiC = 0.34, 1, 2, 3, 4	• The process was found to be less effective for magnetic workpieces. • Ra starts to diminish at a higher speed of rotation due to a decrease in viscosity of media. • The final surface roughness of magnitude 110 nm and 50 nm obtained for SS and brass respectively.
Rotational Assistance				
Sankar et al. (2009)	R-AFF	Al/SiC MMC	• Rotation speed (rpm) = 2–10	• R-AFF produced 44% better surface finish and 81.8% more MR compared to simple AFM process.
Sankar et al. (2016)	R-AFF	Steel 4340	• Workpiece rotational speed (rpm) = 2.0, 4.0, 6.0, 8.0, 10.0	• The higher concentration of abrasive in media results in a decrease in out of roundness (OOR) due to a decrease in flowability of media and self-collision between abrasives, which makes them blunt. • The maximum change in OOR of 39% was observed during the process.
Brar et al. (2013)	HLXAFM	Yellow brass	• With and without a drill	• HLX-AFM resulted in 2.5 times more MR compared to AFM.

(*Continued*)

TABLE 5.1. ***(Continued)***

Experimental Investigations Performed on the Developed Variants of AFM

Reference	Variant	Work Material	Variant-Based Parameter	Variant Performance
			Other Hybrid Variants	
Brar et al. (2015)	ECAFM	Brass	• Media—Polyborosiloxane + hydrocarbon gel + NaI salt + Al2O3 • Voltage (V) = 0, 7, 15 • Salt molal concentration = 0.75, 1, 1.25 • Diameter of cathode rod (mm) = 3.3, 4.2, 5.2	• Electrolytic salts added to media for electrochemical effect. • Optimal MR of 18.02 mg and 46.83% ΔRa was obtained with 0.84 μm final roughness. • Use is limited to the prismatic workpiece due to cathode requirements.
Dabrowski et al. (2006)	ECAFM	Stainless steel	• Voltage (V) = 5–10 • Current (ampere) = 2–8 • Time (min.) = 5–10	• Material removal takes place due to anodic dissolution and abrasion due to abrasives. • Highest MR was reported for cyanide-based electrolytic paste. • The final surface roughness of 0.37 μm was obtained.
Vaishya et al. (2015)	EC2A2FM	Brass	• *Voltage (V) = 0, 15, 25* • *Salt molal concentration = 0.75, 1, 1.25* • *Diameter of cathode rod (mm) = 3.2, 4.2, 5.2* • *rpm of rod = 0, 25, 50* • *abrasive mesh size = 60, 100, 200*	• *A reduction of 70%–80% in machining time was observed.* • *ECM-based contribution was less but significant.*
Ali et al. (2020)	*TACAFM*	*Brass*	• *Current (ampere) = 0, 4, 8, 12, 16* • *Rotation (rpm) = 100, 150, 200, 250, 300* • *Duty cycle = 0.63, 0.68, 0.73, 0.78, 0.83*	• *TACAFM process resulted in 44.34% improvement in MR and 39.74% improvement in Ra compared to conventional AFM.* • *Material removed due to both thermal effect and mechanical abrasion.* • *At higher current (more than 8 ampere), the surface quality deteriorates due to the formation of larger craters on the work surface.*

TABLE 5.2
Detailed Literature about the Performance Analysis of Different Abrasive Media Used in AFM

	Media Constituent			
Inference Drawn	**Base Material**	**Processing Oil**	**Abrasive and Additives**	**Inference Drawn**
			Polymer-Based Abrasive Media	
Bremerstein et al. (2015)	Polyboro-siloxane	Hydrocarbon oil	SiC	• The viscosity and elasticity of media rise with the usage of abrasive media. • No chemical change in media after machining was observed. • A decrease in efficiency of abrasive media was observed for both MR and Ra.
Davies and Fletcher (1995)	Polyboro-siloxane		SiC	• High viscosity media resulted in less temperature rise during the finishing for a fixed finishing time. • Pressure drop and processing time were found to be high for high viscosity media compared to low viscosity media.
Tzeng et al. (2007)	Polymer, wax	Silicone oil	SiC	• The increase in machining time results in a decrease in media viscosity, which allows it to flow through the micro channels. • Decreased viscosity of media helps in finishing of micro holes.
Hull et al. (1992)	Polyboro-siloxane		Diamond and SiC	• At low strain rates, the media behave as apparently dilatant and shows shear thickening behavior. • At high strain rates, media behaves as pseudoplastic and shows shear thinning behavior. • At intermedia strain, rated media behave as a Newtonian fluid. • The stick-slip phenomenon is apparent and significant at low strain rates.
			Magnetorheological Polishing Media	
Jha and Jain (2009)	Grease	Parafin liquid heavy	SiC and CIP	• The performance of media found to increase with an increase in magnetic flux density. • Shear thinning phenomenon was more dominant at the high magnetic field and for coarser grains.

(Continued)

TABLE 5.2. *(Continued)*
Experimental Investigations Performed on the Developed Variants of AFM

	Media Constituent			
Inference Drawn	**Base Material**	**Processing Oil**	**Abrasive and Additives**	**Inference Drawn**
Ghadikolaei and Vahdati (2015)	Glycerin	Liquid paraffin	SiC and iron particle	• Developed media for MRAFF process. • Increase in mesh size results in the better surface quality of work surface due to a decrease in particle size and increased number of cutting edges.
Rubber-Based Abrasive Media				
Sankar et al. (2011)	SBR	Naphthenic oil	SiC	• Media with higher elastic component have a higher radial force and higher MR. • Media with increased stress relaxation have increased elastic nature and have higher MR. • MR was also high for higher storage modulus.
(Kar et al., 2009a)	NR and IIR	Naphthenic oil	SiC	• IIR based abrasive media have better performance compared to NR based media. • Media lose its flow properties with increase in abrasive concentration, and polymer loses its binding properties. • Higher oil concentration resulted in reduced ΔRa.
Wang and Weng (2007)	Silicone rubber	Silicone oil	SiC	• Developed two different media without and with additives. • Media with additives was found to have better physical and thermal properties. • Abrasives with large size result in the better surface finish but consume more time. • Abrasive media only develop a small grinding force, so this process can't remove deep marks.

have been polished through this method (Zhu and Beaucamp, 2020). Initially, the process was deployed to prepare the optical lenses and used to remove the burrs from the silicon wafers (Pal et al., 2017). Later, the process is extensively utilized for semiconductor applications. The schematic diagram and type of material removal mechanism of the CMP process is shown through Figure 5.7 and Figure 5.8 respectively.

The chemical mechanical process is a hybrid process involving both chemical reactions and mechanical action. The material removal is dominated by chemical reactions rather mechanical forces. However, the material removal is mainly governed through Preston equation. Various theories and models have been developed to calculate the

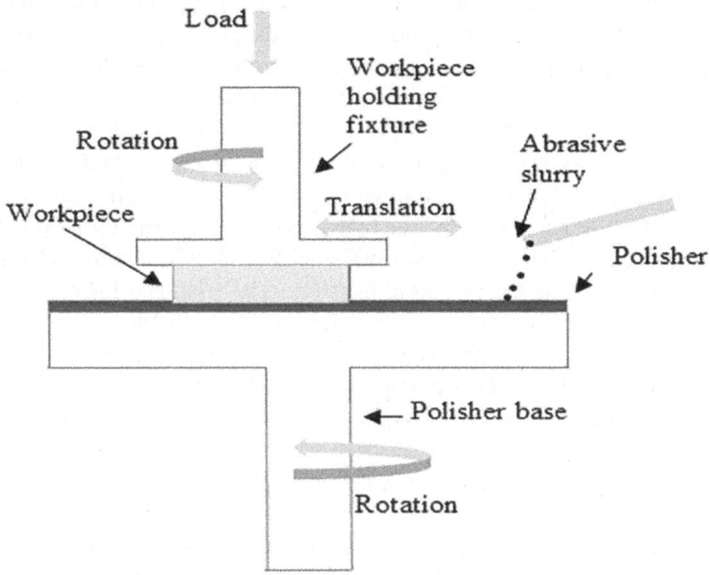

FIGURE 5.7 Experimental setup for full-aperture optical polishing.

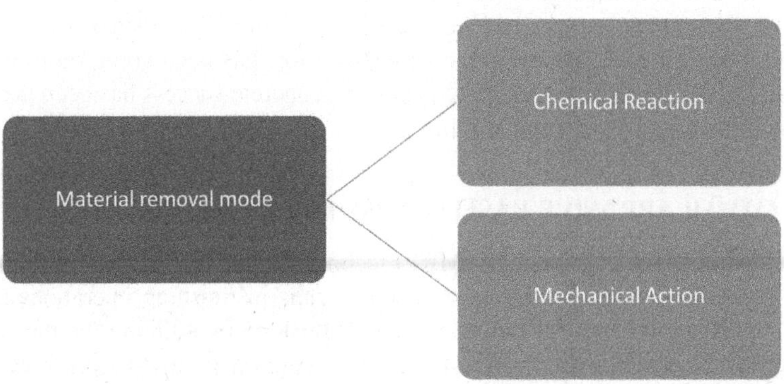

FIGURE 5.8 Different mode of material removal mechanism.

material removal rate (MRR). The Preston's equation used to determine MRR are a base for multiple models. The standard equation of Preston's equation for the material removal rate is given by Eq. 5.1 (Suratwala et al., 2014).

$$MRR = kPV_r \quad (5.1)$$

Where k is Preston's constant, P is the pressure applied, and V_r is the average relative velocity of the polishing particle for the substrate.

In the complete aperture polishing system, an abrasive slurry is continually delivered at the interface as the workpiece rotates against a rotating polisher (a pitch layer or polyurethane pad). The polishing medium, which is a blend of abrasives, is essential for removing the substance. The surface is prepared for polishing by chemical interactions between the workpiece and the polisher at their interface. The workpiece-polisher interface also experiences the effects of touch and lubrication in addition to this chemical interaction. The free abrasives are held by the asperities on the polisher's surface, which prevents them from dispersing when polishing. Between the workpiece and polisher surface, there are two-body contacts that cause physical material removal and three-body contacts that restrict the movement of abrasive particles. The abrasives function as adaptable micro-cutting instruments that remove very little material. The usual load that is delivered to the workpiece disperses unevenly along the polishing interface. The frictional force produced by this normal load and the polisher's rotation aids in material removal.

The CMP process is widely used in semiconductor chip fabrication for various industrial applications. This poses a significant challenge to the environment by increasing greenhouse gases (GHG) due to the excessive use of chemical-based materials and utensils. The components like the polishing pad and abrasive slurry/media are of great concern to review. Chemical disposal and post-cleaning operations significantly impact the ecology and environment, as most are nonreusable. Minimal research on sustainability, life cycle assessment, and post-cleaning methods are available. The energy analysis for the CMP process and its consumables includes raw materials, transportation, electricity, water purifications, and post-cleaning. The different CMP-related chemicals and wastewater-cleaning materials are shown in Figure 5.9. The electrocoagulation process is one of the most influential and cost-effective ways to remove the particles from wastewater. The treated water can be used for reuse for different uses. A continuous effort has been going on to recover the abrasive particles. However, there is no very concrete success has been achieved in this research area (Lee et al., 2021a).

5.5 OTHER ABRASIVE BASED FINISHING OPERATIONS

The magnetorheological finishing (MRF) technique, essentially a sub-aperture finishing procedure, is one of the most current trends in finishing operations (Singh et al., 2020b). It applies a regulated amount of finishing by utilizing the rheological qualities of the abrasive slurry. The material is removed from the workpiece using force. According to Jacobs et al. (1995), the procedure involves using magnetorheological (MR) fluid as a slurry that contains carbonyl iron particles, abrasives, and

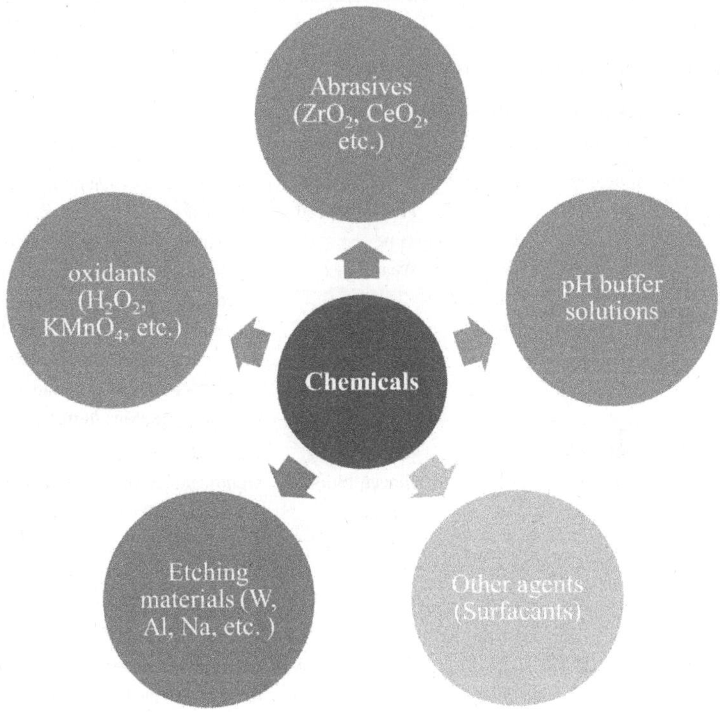

FIGURE 5.9 Chemicals in the post-cleaning methods.

liquid media, such as water, oil, grease, etc. By adding a magnetic field, one may adjust the softening or hardening of the MR fluid due to the presence of carbonyl iron particles. This idea underpins the MRF polishing process. When in contact with the workpiece under a magnetic field, MR fluid becomes stiffened and works as a cutting tool. Abrasive particles are held toward the workpiece surface, while carbonyl iron particles align in a specific direction in a magnetic field. This process tends to be deterministic since the finishing forces can be controlled by manipulating the strength of the MR fluid by creating a magnetic field (Sidpara and Jain, 2012). Although the procedure seems promising, specific problems with MR fluid stability, the choice of MR fluid parameters for various materials, etc., limit its current application. The method can make spot repairs after complete aperture polishing since it has a low material removal rate (MRR) and can remove material from the workpiece one area at a time.

The abrasive mixed viscous base medium overcomes the shape limitation in practically all conventional finishing methods by acting as a self-deforming stone. As was previously mentioned, the AFM process lacks determinism because the abrasive forces are the least controllable by outside influences. A new hybrid method known as magnetorheological abrasive flow finishing (MRAFF) has been created to maintain the AFM process's flexibility. The process also having determinism and controllability of the rheological features of the abrasive-laden media (Jha and Jain, 2004). An

TABLE 5.3
Key Research Articles in Chemical Mechanical Polishing

Authors	Workpiece Material	Abrasive Powder	Input Parameters	Inference
Pal et al. (2018)	BK7 optical glass	Cerium oxide (CeO_2) powder (Rhodite 200)	Normal load Polisher speed Time Wear index	The friction coefficient determines the nature of material removal. At high-speed ductile mode, ductile cum brittle mode of material removal mechanism was observed at the low speed irrespective of the load. Further, the friction coefficient suggests mechanical action is dominant over chemical reactions during polishing.
Pal et al. (2016)	BK7 optical glass	CeO_2 powder	Abrasive concentration Pressure Overarm speed	Abrasive concentration is the most significant parameter among input parameters. The chemical mode of material removal ensures no redisposition of removed material forms over the polished surface by the dissolution process.
(Singh et al., 2017a)	BK7 optical glass 12 mm thick	CeO_2 powder 5% wt./vol	Relative velocity Normal load Abrasive size	Polishing kinematics greatly influence the surface finish of the polished surface. The relative velocity is the most significant parameter, while the average load affects the least. Small-size abrasive particles yield in better surface finish compared to a larger one.
Kumar and Omkumar (2019) (Ramesh Kumar and Omkumar, 2019)	Soda-lime glass	CeO_2 powder	Flow rate Speed Down pressure Processing	Speed and down pressure have the maximum influence over the material removal rate and surface roughness, respectively. The 0.9924 mm2/min and 0.3925 μm are the best optimal response values obtained using the hybrid Taguchi and ANOVA methodology.
(Xue et al., 2019)	Chalcogenide glasses Ge10As40Se50 (IG-4),	Alumina (Al_2O_3) 1.0 μm	Slurry composition	The surface roughness can be improved with NaOH and H_2O_2 chemicals. The surface defects are significantly reduced at higher CMP slurry concentrations (H_2O_2). The best surface roughness was obtained at 0.292 nm at 20% wt. of H_2O_2.
(Singh et al., 2017b)	Fused silica glass samples 100 mm	CeO_2 powder	Abrasive size Concentration Relative velocity	The relative velocity and its interaction with concentration are found to be significant. The abrasion theory explains the results obtained, but the chemical aspect cannot be ignored. The best surface roughness was received at 7 nm.

MR-polishing fluid used in the MRAFF process has rheological characteristics that an external magnetic field can control. This method can smooth out complex interior and external geometries with surface roughness values as low as nanometers.

The spherical shaped components are finished through a differently developed process called magnetic flow polishing (MFP). The spherical shaped parts are not processed through the aforementioned finishing techniques. In this process, magnetic field given to the abrasive slurry is utilized to finish the spherical components.

5.6 RESEARCH CHALLENGES OF ADVANCED ABRASIVE-BASED FINISHING PROCESSES

The advanced abrasive-based finishing processes are effective nano-finishing techniques; nevertheless, there are several unresolved issues. The phenomenon, cause, and proposed solutions of various research challenges of advanced abrasive-based finishing processes have been discussed in this section. In the AFM process, the finishing time is high. Externally viscosity of AFM media cannot be controlled. Higher viscosity media are used where the passage is larger and where the passage is smaller or around radius edges; lower viscosity media are used. Media viscosity changes with change in shear rate, shear stress, media temperature, grain diameter, base medium, or by changing the percentage concentration of abrasives. MRR and surface quality improves with increased medium viscosity (Loveless et al., 1994). The AFM process is not able to properly finish the blind holes, and also, its setup and media cost needs high capital investment. The cost of viscoelastic carriers like silicon or polyborosilixane (PBS) is too high, which makes the whole AFM media highly costly. In place of these carrier media, natural rubber or butyl rubber media can be used, which is easily available and has a low cost (Kar et al., 2009a). For controlling the viscosity of AFM media, the proper percentage of the viscoelastic carrier should be mixed with processing oil (turpentine oil); it is generally kept in between 2.5:1 ratio by weight. The weight ratio of carrier media and abrasives also should be between 1:1 to 1.5:1 for proper maintenance of viscosity of media (Jain and Adsul, 2000). By optimizing the process parameters, the finishing time of the AFM process can be reduced (Cheema et al., 2012). In MFAF processes, finishing time is also high. The MFAF process, like MAF, can be difficult to implement in mass production operations due to high finishing time (Mikno, 2018). In this MAF process, friction between magnetic abrasive particles (MAPs) and walls of the container causes structural changes in IPs. This changed IPs structure decreases the finishing performance due to a decrease in their magnetization ability causing a reduction in the holding capacity of abrasives. The higher rotational speed of the roller workpiece also reduces the finishing ability of MAF processing (Fox et al., 1994). As the rotational speed of the workpiece increases, there will be less time available for the contact of MAPs with the roller surface, which causes inappropriate finishing. The optimum rotational speed will maintain the finishing efficiency and time. The other MFAF process, like MFP, is mainly used for spherical ball polishing.

Certain modification in the MFP setup is needed for making it flexible for proper polishing of other complex components. In place of vertical plane drive shafts, free-form shafts can be used, which will be the same as the shape of the workpiece

(Umehara et al., 2006). This modification in the setup can make the MFP process more flexible. MRFF processes take a higher amount of finishing time to polish tough material; however, it takes less time to polish soft materials. High-quality MRPFs are expensive (Jacobs et al., 1995; Golini et al., 1998). The machining setup of the MRF process is complex for freeform optics polishing. Settling of ferromagnetic IPs in this MRF process is also a major problem. Stability issues are also observed in the MRF processing. Stability is the resilience of MRPFs to endure sedimentation and agglomeration of ferromagnetic particles. The suspension of the ferromagnetic particles stays homogenous in the MRPF system, which keeps sedimentation stability consistent. In the lack of a magnetic field, the capacity of the MRPF to stay in a distributed state rather than agglomeration is referred to as agglomerative stability. An additive is often used to improve sedimentation stability, such as when stearic acid is added (Barman and Das, 2018). The presence of iron naphthalate and iron stearate increased the dispersibility of IPs (Nagdeve et al., 2018b). By introducing nano-sized IPs with the micro-sized IPs in MRPF, the yield stress and the sedimentation stability of the MRPF are improved. A common method of controlling sedimentation is to use thixotropic agents and surfactants, such as silica gel, stearates, xanthan gum, grease, and carboxylic acids. When flow is disrupted at a lower shear rate, the thixotropic network shows excellent viscosity, but when the shear rate is raised, the network becomes thinner. In addition, treating the polishing particles with polymers may help to enhance the dispersion stability of the MRPF particles (Kumar et al., 2019). In the MRF process, relatively higher power is required. This process is generally not preferable for polishing the inner surface/outer complex contour of the free-form component (Yang et al., 2017). In the CMRF process, non-uniform corrosion layers are formed on SiC wafers at higher concentrations of IPs, which ultimately reduces the finishing performance (Liang et al., 2016). The proper use of a higher concentration of chemicals in CMRF polishing media damages the workpiece surface and sometimes causes burning along the workpiece surfaces (Liang et al., 2018). The handling of chemicals in the CMRF polishing media is also a major problem (Kumari and Chak, 2018). During CMRF processing, a caring environment is needed, where the operator has to definitely use safety equipment. In the BEMRF process, continuous replenishment of MRPF is needed at the cylindrical tool end (Saraswathamma et al., 2015). If MRPF at the tool end is not replaced, then abrasives of MRPF get blunt, and it will not properly polish the workpiece surface after a certain time. For solving this issue, a throughout circular hole can be made at the center inside the BEMRF tool, where a pipe can be connected to continuously flow the MPF from the media container. The heating of the electromagnet coil in the BEMRF processes causes a decrement in viscosity of MRPF, as a consequence of which, the inappropriate development of a polishing brush at the end of the cylindrical tool takes place (Khan and Jha, 2019). This issue could be solved through the usage of a permanent magnet, where before using it, its magnetic flux density distribution can be optimized using FEA based software like Ansys® Maxwell. Also, to ensure that only polishing brush form at the end of the BEMRF tool, a proper magnetic shielding material like Mu-metal can be used to make the tool. Because of the spatial variation in magnetic field distributions (MFDs) along workpiece surfaces, uniform polishing of free-form surfaces is difficult to achieve in the MRAFF method [44]. This spatial variation of MFDs is caused by the positions of magnets from workpiece freeform surfaces, as the distance of each magnet is at a different distance from each point on the workpiece surface. To avoid this issue, a negative replica of workpiece profiles can be used as a fixture in the MRAFF method (Barman and Das, 2017). This fixture will force the MRPF to uniformly pass across each and every corner of the workpiece profiles for uniform finishing. After a large number of finishing cycles, sedimentation issue in MRPF during the MRAFF process occurs due to frictional heat generation, which decreases the viscosity of MRPF (Kanthale and Pande, 2019). MRPF must be replaced after a certain amount of time. One of the other issues is particle clogging in experimental setup fixtures when subjected to a non-uniform magnetic field and instability in applied pressure. To avoid this issue, some additives can be applied to MRPF (Jain, 2008). During the MRAFF method of polishing hard materials, it was discovered that certain unnecessary shape modifications occur mostly at the workpiece surface's corner points and sharp edges (Houshi, 2016). This may be attributed to instabilities in applied pressure during the MRPF's continuous to-and-fro movements. The MRAFF method often has difficulty in polishing spherical components due to its free-from surfaces, which needs a complex fixture design. To a certain extent, the composition of the MRPF shifts due to the mixing of the workpiece's induced nano chips during the MRAFF process. MRAFF experiment setup is tedious in nature; due to this, sometimes leakage problem from the experimental setup occurs. For overcoming this issue, proper sealing materials like rubber or gasket can be used (Nagdeve et al., 2018a). In the R-MRAFF process, the off-state viscosity of MRPF increases at higher finishing cycles. As the MRPF's off-state viscosity increases, radial force and yield shear stress drop as well reduce the MRR and surface finishing rate. One of the major problems in the R-MRAFF and MRAFF process, for every differently shaped complex component, a different fixture is needed. Higher rotational speed causes more vibration to the R-MRAFF setup reducing finishing performance (Kanthale and Pande, 2019). The higher rotational speed of magnets/workpiece causes the generation of large centrifugation force and tangential force, which makes the abrasives break the IPs chain and move freely without any polishing activity. An optimum rotational speed is needed for maximizing the finishing performance. The optimum rotational speed of magnets/workpieces varies from 80–100 rpm (Das et al., 2010; Das et al., 2012b). Further, many other studies reported the post-processing over additive manufactured substrates. For getting finished surfaces, several finishing and machining processes have been implemented. In addition, while most post-processing procedures involve a single finishing step, AM components can be completed with hybrid successive operations, allowing manufacturers to make use of a variety of post-processing methods without being constrained by any one of them.

5.7 CONCLUSIONS

In this article, the comprehensive discussion has been made to understand the abrasive-based finishing processes. The abrasive particle majorly plays an important role in effective removal of material from the surface during finishing along with some external actions like magnetic forces, mechanical forces, ultrasonic vibrations, etc.

The past research conducted in this domain has been discussed. The research gaps have been drawn to explore the potential research areas. The following conclusions may be drawn from the previous discussion.

- The primary manufacturing processes yield surface irregularities. The conventional finishing is not able to finish the surface and remove all defects including sub-aperture disparities.
- The advanced finishing can finish the surface with required accuracy and precision at the micro and nano level.
- The AFM and its variants can achieve better surface compared to conventional finishing operations with lower cost by controlling the process. The process has potential applications in miniature applications such as MEMS/MOEMS components.
- The magnetic field plays an important role in directing the abrasive cutting forces during finishing at the micro and nano level. The pulsating effect DC power and magnetic forces (electromagnet) considerably improve the surface finish.
- Frictional force has an essential parameter in governing the material removal rate (MRR) and surface integrity parameters. The CMP performance is a semimanual skilled process, and finishing highly depends on the handling of the machine.
- The disposal of abrasive media and slurry have a critical consequence to the operator and environment. There is very limited research available on the sustainability of the processes.

REFERENCES

Ali, P., Walia, R., Murtaza, Q., & Singari, R. M. (2020). Material removal analysis of hybrid EDM-assisted centrifugal abrasive flow machining process for performance enhancement. *Journal of the Brazilian Society of Mechanical Sciences and Engineering*, *42*(6), 1–28. https://doi.org/10.1007/s40430-020-02375-6

Allen, D. (2003). The state of the art of photochemical machining at the start of the twenty-first century. *Proceedings of the Institution of Mechanical Engineers, Part B: Journal of Engineering Manufacture*, *217*(5), 643–650. https://doi.org/10.1243/095440503322011362

Annamaria, G., Massimiliano, B., & Francesco, V. (2022). Laser polishing: A review of a constantly growing technology in the surface finishing of components made by additive manufacturing. *The International Journal of Advanced Manufacturing Technology*, 1–40. https://doi.org/10.1007/s00170-022-08840-x

Babbar, A., Jain, V., & Gupta, D. (2020). In vivo evaluation of machining forces, torque, and bone quality during skull bone grinding. *Proceedings of the Institution of Mechanical Engineers, Part H: Journal of Engineering in Medicine*, *234*(6), 626–638. https://doi.org/10.1177/0954411920911499

Babbar, A., Jain, V., Gupta, D., & Agrawal, D. (2021a). Finite element simulation and integration of CEM43 °C and Arrhenius Models for ultrasonic-assisted skull bone grinding: A thermal dose model. *Medical Engineering and Physics*, *90*, 9–22. https://doi.org/10.1016/j.medengphy.2021.01.008

Babbar, A., Jain, V., Gupta, D., & Agrawal, D. (2021b). Histological evaluation of thermal damage to Osteocytes: A comparative study of conventional and ultrasonic-assisted bone grinding. *Medical Engineering and Physics*, *90*, 1–8. https://doi.org/10.1016/j.medengphy.2021.01.009

Babbar, A., Jain, V., Gupta, D., Agrawal, D., Prakash, C., Singh, S., Wu, L. Y., Zheng, H. Y., Królczyk, G., & Bogdan-Chudy, M. (2021c). Experimental analysis of wear and multi-shape burr loading during neurosurgical bone grinding. *Journal of Materials Research and Technology*, *12*, 15–28. https://doi.org/10.1016/j.jmrt.2021.02.060

Babbar, A., Jain, V., Gupta, D., Prakash, C., & Agrawal, D. (2022a). Potential application of CEM43° C and Arrhenius model in neurosurgical bone grinding. In *Numerical Modelling and Optimization in Advanced Manufacturing Processes* (pp. 145–158). https://doi.org/10.1007/978-3-031-04301-7_9

Babbar, A., Sharma, A., & Singh, P. (2022b). Multi-objective optimization of magnetic abrasive finishing using grey relational analysis. *Materials Today: Proceedings*, *50*, 570–575. https://doi.org/10.1016/j.matpr.2021.01.004

Barman, A., & Das, M. (2017). Design and fabrication of a novel polishing tool for finishing freeform surfaces in magnetic field assisted finishing (MFAF) process. *Precision Engineering*, *49*, 61–68. https://doi.org/10.1016/j.precisioneng.2017.01.010

Barman, A., & Das, M. (2018). Nano-finishing of bio-titanium alloy to generate different surface morphologies by changing magnetorheological polishing fluid compositions. *Precision Engineering*, *51*, 145–152. https://doi.org/10.1016/j.precisioneng.2017.08.003

Brar, B., Walia, R., & Singh, V. (2015). Electrochemical-aided abrasive flow machining (ECA2FM) process: A hybrid machining process. *The International Journal of Advanced Manufacturing Technology*, *79*(1), 329–342. https://doi.org/10.1007/s00170-015-6806-y

Brar, B., Walia, R., Singh, V., & Sharma, M. (2013). A robust helical abrasive flow machining (HLX-AFM) process. *Journal of the Institution of Engineers (India): Series C*, *94*(1), 21–29. https://doi.org/10.1007/s40032-012-0054-9

Bremerstein, T., Potthoff, A., Michaelis, A., Schmiedel, C., Uhlmann, E., Blug, B., & Amann, T. (2015). Wear of abrasive media and its effect on abrasive flow machining results. *Wear*, *342*, 44–51. http://doi.org/10.1016/j.wear.2015.08.013

Cheema, M. S., Venkatesh, G., Dvivedi, A., & Sharma, A. K. (2012). Developments in abrasive flow machining: A review on experimental investigations using abrasive flow machining variants and media. *Proceedings of the Institution of Mechanical Engineers, Part B: Journal of Engineering Manufacture*, *226*(12), 1951–1962. https://doi.org/10.1177/0954405412462000

Dabrowski, L., Marciniak, M., & Szewczyk, T. (2006). Analysis of abrasive flow machining with an electrochemical process aid. *Proceedings of the Institution of Mechanical Engineers, Part B: Journal of Engineering Manufacture*, *220*(3), 397–403. https://doi.org/10.1243/095440506X77571

Das, M., Jain, V., & Ghoshdastidar, P. (2010). Nano-finishing of stainless-steel tubes using rotational magnetorheological abrasive flow finishing process. *Machining Science and Technology*, *14*(3), 365–389. https://doi.org/10.1080/10910344.2010.511865

Das, M., Jain, V., & Ghoshdastidar, P. (2012a). Nanofinishing of flat workpieces using rotational—magnetorheological abrasive flow finishing (R-MRAFF) process. *The International Journal of Advanced Manufacturing Technology*, *62*(1), 405–420. https://doi.org/10.1007/s00170-011-3808-2

Das, M., Jain, V., & Ghoshdastidar, P. (2012b). Computational fluid dynamics simulation and experimental investigations into the magnetic-field-assisted nano-finishing process.

Proceedings of the Institution of Mechanical Engineers, Part B: Journal of Engineering Manufacture, *226*(7), 1143–1158. https://doi.org/10.1177/0954406211426948

Davies, P., & Fletcher, A. (1995). The assessment of the rheological characteristics of various polyborosiloxane/grit mixtures as utilized in the abrasive flow machining process. *Proceedings of the Institution of Mechanical Engineers, Part C: Journal of Mechanical Engineering Science*, *209*(6), 409–418. https://doi.org/10.1243/PIME_PROC_1995_209_171_02

Debnath, S., Kunar, S., Anasane, S., & Bhattacharyya, B. (2017). Non-traditional micromachining processes: Opportunities and challenges. *Non-traditional Micromachining Processes*, 1–59. https://doi.org/10.1007/978-3-319-52009-4_1

Dehghan Ghadikolaei, A., & Vahdati, M. (2015). Experimental study on the effect of finishing parameters on surface roughness in magneto-rheological abrasive flow finishing process. *Proceedings of the Institution of Mechanical Engineers, Part B: Journal of Engineering Manufacture*, *229*(9), 1517–1524. https://doi.org/10.1177/0954405414539488

Dixit, N., Sharma, V., & Kumar, P. (2021a). Experimental investigations into abrasive flow machining (AFM) of 3D printed ABS and PLA parts. *Rapid Prototyping Journal*, *28*(1), 161–174. https://doi.org/10.1108/RPJ-01-2021-0013

Dixit, N., Sharma, V., & Kumar, P. (2021b). Research trends in abrasive flow machining: A systematic review. *Journal of Manufacturing Processes*, *64*, 1434–1461. https://doi.org/10.1016/j.jmapro.2021.03.009

Dubey, A. K., & Yadava, V. (2008). Laser beam machining—A review. *International Journal of Machine Tools and Manufacture*, *48*(6), 609–628. https://doi.org/10.1016/j.ijmachtools.2007.10.017

Fang, F., Zhang, X., Weckenmann, A., Zhang, G., & Evans, C. (2013). Manufacturing and measurement of freeform optics. *CIRP Annals*, *62*(2), 823–846. https://doi.org/10.1016/j.cirp.2013.05.003

Fang, L., Zhao, J., Sun, K., Zheng, D., & Ma, D. (2009). Temperature as sensitive monitor for efficiency of work in abrasive flow machining. *Wear*, *266*(7–8), 678–687. https://doi.org/10.1016/j.wear.2008.08.014

Fox, M., Agrawal, K., Shinmura, T., & Komanduri, R. (1994). Magnetic abrasive finishing of rollers. *CIRP Annals*, *43*(1), 181–184. https://doi.org/10.1016/S0007-8506(07)62191-X

Golini, D., Dumas, P., Kordonski, W., Hogan, S., & Jacobs, S. (1998). *Precision Optics Fabrication Using Magnetorheological Finishing*. Paper Presented at the Optical Fabrication and Testing. https://doi.org/10.1117/12.279809

Gudipadu, V., Sharma, A. K., & Singh, N. (2015). Simulation of media behaviour in vibration assisted abrasive flow machining. *Simulation Modelling Practice and Theory*, *51*, 1–13. https://doi.org/10.1016/j.simpat.2014.10.009

Hassanin, H., Elshaer, A., Benhadj-Djilali, R., Modica, F., & Fassi, I. (2018). Surface finish improvement of additive manufactured metal parts. *Micro and Precision Manufacturing*, 145–164. http://doi.org/10.1007/978-3-319-68801-5

Heng, L., Kim, J. S., Song, J. H., & Mun, S. D. (2022). A review on surface finishing techniques for difficult-to-machine ceramics by non-conventional finishing processes. *Materials*, *15*(3), 1227. https://doi.org/10.3390/ma15031227

Houshi, M. N. (2016). *A Comprehensive Review on Magnetic Abrasive Finishing Process*. Paper Presented at the Advanced Engineering Forum, 18, 1–20. https://doi.org/10.4028/www.scientific.net/AEF.18.1

Hull, J., O'Sullivan, D., Fletcher, A., Trengove, S., & Mackie, J. (1992). Rheology of carrier media used in abrasive flow machining. *Key Engineering Materials*, *72*, 617–626.

Jacobs, S. D., Golini, D., Hsu, Y., Puchebner, B. E., Strafford, D., Prokhorov, I. V., . . . Kordonski, W. I. (1995). *Magnetorheological Finishing: A Deterministic Process for*

Optics Manufacturing. Paper Presented at the International Conference on Optical Fabrication and Testing. https://doi.org/10.1117/12.215617

Jain, V. (2008). Abrasive-based nano-finishing techniques: An overview. *Machining Science and Technology*, *12*(3), 257–294. https://doi.org/10.1080/10910340802278133

Jain, V., & Adsul, S. (2000). Experimental investigations into abrasive flow machining (AFM). *International Journal of Machine Tools and Manufacture*, *40*(7), 1003–1021. https://doi.org/10.1016/S0890-6955(99)00114-5

Jain, V., Ranganatha, C., & Muralidhar, K. (2001). Evaluation of rheological properties of medium for AFM process. *Machining Science and Technology*, *5*, 151–170. https://doi.org/10.1081/MST-100107841

Jefferies, S. R. (2007). Abrasive finishing and polishing in restorative dentistry: A state-of-the-art review. *Dental Clinics of North America*, *51*(2), 379–397. https://doi.org/10.1016/j.cden.2006.12.002

Jha, S., & Jain, V. (2004). Design and development of the magnetorheological abrasive flow finishing (MRAFF) process. *International Journal of Machine Tools and Manufacture*, *44*(10), 1019–1029. https://doi.org/10.1016/j.ijmachtools.2004.03.007

Jha, S., & Jain, V. (2009). Rheological characterization of magnetorheological polishing fluid for MRAFF. *The International Journal of Advanced Manufacturing Technology*, *42*(7), 656–668. https://doi.org/10.1007/s00170-008-1637-8

Jones, A., & Hull, J. (1998). Ultrasonic flow polishing. *Ultrasonics*, *36*(1–5), 97–101. https://doi.org/10.1016/S0041-624X(97)00147-9

Kalisz, J., Żak, K., Wojciechowski, S., Gupta, M., & Krolczyk, G. (2021). Technological and tribological aspects of milling-burnishing process of complex surfaces. *Tribology International*, *155*, 106770. https://doi.org/10.1016/j.triboint.2020.106770

Kanthale, V., & Pande, D. (2019). Experimental characterization of surface roughness using magnetorheological abrasive flow finishing process on AISI D3 steel. *Journal of Bio- and Tribo-Corrosion*, *5*(3), 1–16. https://doi.org/10.1007/s40735-019-0254-4

Kar, K. K., Ravikumar, N., Tailor, P. B., Ramkumar, J., & Sathiyamoorthy, D. (2009a). Performance evaluation and rheological characterization of newly developed butyl rubber based media for abrasive flow machining process. *Journal of Materials Processing Technology*, *209*(4), 2212–2221. https://doi.org/10.1016/j.jmatprotec.2008.05.012

Kar, K. K., Ravikumar, N., Tailor, P. B., Ramkumar, J., & Sathiyamoorthy, D. (2009b). Preferential media for abrasive flow machining. *Journal of Manufacturing Science and Engineering*, *131*(1). https://doi.org/10.1115/1.3046135

Khan, D. A., & Jha, S. (2019). Selection of optimum polishing fluid composition for ball end magnetorheological finishing (BEMRF) of copper. *The International Journal of Advanced Manufacturing Technology*, *100*(5), 1093–1103. https://doi.org/10.1007/s00170-017-1056-9

Khatri, N., Manoj, J. X., Mishra, V., Garg, H., & Karar, V. (2018). Experimental and simulation study of nanometric surface roughness generated during magnetorheological finishing of silicon. *Materials Today: Proceedings*, *5*(2), 6391–6400. https://doi.org/10.1016/j.matpr.2017.12.250

Kumar, J. S., Paul, P. S., Raghunathan, G., & Alex, D. G. (2019). A review of challenges and solutions in the preparation and use of magnetorheological fluids. *International Journal of Mechanical and Materials Engineering*, *14*(1), 1–18. https://doi.org/10.1186/s40712-019-0109-2

Kumar, M., Alok, A., Kumar, V., & Das, M. (2022a). Advanced abrasive-based nano-finishing processes: Challenges, principles and recent applications. *Materials and Manufacturing Processes*, *37*(4), 372–392. https://doi.org/10.1080/10426914.2021.2001509

Kumar, M., Oza, A. D., Prajapati, M., & Joshi, G. (2020a). *Experimental Investigation During Machining of P20 Tool Steel Using EDM*. Paper Presented at the International Conference on Advances in Materials Processing & Manufacturing Applications. https://doi.org/10.1007/978-981-16-0909-1_56

Kumar, M., & Satsangi, P. (2021). A study on machining performance of wire electric discharge grinding (WEDG) process during machining of tungsten alloy micro-tools. *Sādhanā*, *46*(2), 1–11. https://doi.org/10.1007/s12046-021-01595-3

Kumar, M., Singh Yadav, H. N., Kumar, A., & Das, M. (2021a). An overview of magnetorheological polishing fluid applied in nano-finishing of components. *Journal of Micromanufacturing*, 25165984211008173. https://doi.org/10.1177/25165984211008173

Kumar, M., Vaishya, R., Oza, A. D., & Suri, N. M. (2020b). Experimental investigation of wire-electrochemical discharge machining (WECDM) performance characteristics for quartz material. *Silicon*, *12*(9), 2211–2220. https://doi.org/10.1007/s12633-019-00309-z

Kumar, M., Vaishya, R., & Suri, N. (2022b). Simulation and experimental research of material removal rate in micro-electrochemical discharge machining process. *Advances in Materials and Processing Technologies*, 1–16. https://doi.org/10.1080/2374068X.2022.2127936

Kumar, M., Vaishya, R., Suri, N., & Manna, A. (2021b). An experimental investigation of surface characterization for zirconia ceramic using electrochemical discharge machining process. *Arabian Journal for Science and Engineering*, *46*(3), 2269–2281. https://doi.org/10.1007/s13369-020-05059-4

Kumar, Y., & Singh, H. (2022). Chemomechanical magnetorheological finishing: Process mechanism, research trends, challenges and opportunities in surface finishing. *Journal of Micromanufacturing*, *5*(2), 193–206. https://doi.org/10.1177/25165984211038878

Kumari, C., & Chak, S. K. (2018). A review on magnetically assisted abrasive finishing and their critical process parameters. *Manufacturing Review*, *5*, 13. https://doi.org/10.1051/mfreview/2018010

Lee, H., Kim, H., & Jeong, H. (2021a). Approaches to sustainability in chemical mechanical polishing (CMP): A review. *International Journal of Precision Engineering and Manufacturing-Green Technology*, 1–19. https://doi.org/10.1007/s40684-021-00406-8

Lee, J.-Y., Nagalingam, A. P., & Yeo, S. (2021b). A review on the state-of-the-art of surface finishing processes and related ISO/ASTM standards for metal additive manufactured components. *Virtual and Physical Prototyping*, *16*(1), 68–96. https://doi.org/10.1080/17452759.2020.1830346

Liang, H. Z., Lu, J., Pan, J., & Yan, Q. (2018). Material removal process of single-crystal SiC in chemical-magnetorheological compound finishing. *The International Journal of Advanced Manufacturing Technology*, *94*(5), 2939–2948. https://doi.org/10.1007/s00170-017-1098-z

Liang, H. Z., Yan, Q. S., Lu, J. B., & Gao, W. Q. (2016). *Experiment on Chemical Magnetorheological Finishing of SiC Single Crystal Wafer*. Paper Presented at the Materials Science Forum. https://doi.org/10.1007/s00170-019-03594-5

Loveless, T. R., Williams, R., & Rajurkar, K. P. (1994). A study of the effects of abrasive-flow finishing on various machined surfaces. *Journal of Materials Processing Technology*, *47*(1–2), 133–151. https://doi.org/10.1016/0924-0136(94)90091-4

Mikno, Z. (2018). Projection welding of nuts involving the use of electromechanical and pneumatic electrode force. *International Journal of Advanced Manufacturing Technology*, *99*. https://doi.org/10.1007/s00170-018-2525-5

Mizobuchi, A., & Tashima, A. (2020). Optimization of wet grinding conditions of sheets made of stainless steel. *Journal of Manufacturing and Materials Processing*, *4*(4), 114. https://doi.org/10.3390/jmmp4040114

Nagdeve, L., Jain, V., & Ramkumar, J. (2018a). Nanofinishing of freeform/sculptured surfaces: State-of-the-art. *Manufacturing Review, 5*, 6. https://doi.org/10.1051/mfreview/2018005

Nagdeve, L., Sidpara, A., Jain, V., & Ramkumar, J. (2018b). On the effect of relative size of magnetic particles and abrasive particles in MR fluid-based finishing process. *Machining Science and Technology, 22*(3), 493–506. https://doi.org/10.1080/10910344.2017.1365899

Pal, R. K., Garg, H., & Karar, V. (2017). Material removal characteristics of full aperture optical polishing process. *Machining Science and Technology, 21*(4), 493–525. https://doi.org/10.1080/10910344.2017.1336626

Pal, R. K., Garg, H., Sarepaka, R. V., & Karar, V. (2016). Experimental investigation of material removal and surface roughness during optical glass polishing. *Materials and Manufacturing Processes, 31*(12), 1613–1620. https://doi.org/10.1080/10426914.2015.1103867

Pal, R. K., Kumar, M., & Karar, V. (2022b). Experimental investigation of polishing process for Schott BK-7 optical glass. *Materials Today: Proceedings, 57*, 734–738. https://doi.org/10.1016/j.matpr.2022.02.218

Pal, R. K., Sharma, R., Baghel, P. K., Garg, H., & Karar, V. (2018). An approach for quantification of friction and enhancing the process efficiency during polishing of optical glass. *Journal of Mechanical Science and Technology, 32*(8), 3835–3842. https://doi.org/10.1007/s12206-018-0735-2

Prakash, C., Singh, S., Pramanik, A., Basak, A., Królczyk, G., Bogdan-Chudy, M., . . . Zheng, H. (2021). Experimental investigation into nano-finishing of β-TNTZ alloy using magnetorheological fluid magnetic abrasive finishing process for orthopedic applications. *Journal of Materials Research and Technology, 11*, 600–617. https://doi.org/10.1016/j.jmrt.2021.01.046

Ramesh Kumar, C., & Omkumar, M. (2019). Optimisation of process parameters of chemical mechanical polishing of soda lime glass. *Silicon, 11*(1), 407–414. https://doi.org/10.1007/s12633-018-9903-3

Sahu, A. K., Malhotra, J., & Jha, S. (2022). Laser-based hybrid micromachining processes: A review. *Optics & Laser Technology, 146*, 107554. https://doi.org/10.1016/j.optlastec.2021.107554

Sankar, M. R., Jain, V., & Ramkumar, J. (2009). Experimental investigations into rotating workpiece abrasive flow finishing. *Wear, 267*(1–4), 43–51. https://doi.org/10.1016/j.wear.2008.11.007

Sankar, M. R., Jain, V., & Ramkumar, J. (2016). Nano-finishing of cylindrical hard steel tubes using rotational abrasive flow finishing (R-AFF) process. *The International Journal of Advanced Manufacturing Technology, 85*(9), 2179–2187. https://doi.org/10.1007/s00170-015-8189-5

Sankar, M. R., Jain, V., Ramkumar, J., & Joshi, Y. (2011). Rheological characterization of styrene-butadiene based medium and its finishing performance using rotational abrasive flow finishing process. *International Journal of Machine Tools and Manufacture, 51*(12), 947–957. https://doi.org/10.1016/j.ijmachtools.2011.08.012

Saraswathamma, K., Jha, S., & Rao, P. V. (2015). Rheological characterization of MR polishing fluid used for silicon polishing in BEMRF process. *Materials and Manufacturing Processes, 30*(5), 661–668. https://doi.org/10.1080/10426914.2014.994767

Sharma, A., Babbar, A., Tian, Y., Pathri, B. P., Gupta, M., & Singh, R. (2022a). Machining of ceramic materials: A state-of-the-art review. *International Journal on Interactive Design and Manufacturing*. https://doi.org/10.1007/s12008-022-01016-7

Sharma, A., & Jain, V. (2020b). Experimental investigation of cutting temperature during drilling of float glass specimen. *IOP Conference Series: Materials Science and Engineering, 715*(1), 012050. https://doi.org/10.1088/1757-899X/715/1/012050

Sharma, A., Jain, V., Gupta, D., & Babbar, A. (2020). A review study on miniaturization. In *Advanced Manufacturing and Processing Technology* (1st ed., pp. 111–131). Boca Raton. https://doi.org/10.1201/9780429298042-5

Sharma, A., Jain, V., & Gupta, D. (2018). Characterization of chipping and tool wear during drilling of float glass using rotary ultrasonic machining. *Measurement: Journal of the International Measurement Confederation, 128*, 254–263. https://doi.org/10.1016/j.measurement.2018.06.040

Sharma, A., Jain, V., & Gupta, D. (2019a). Multi-shaped tool wear study during rotary ultrasonic drilling and conventional drilling for amorphous solid. *Proceedings of the Institution of Mechanical Engineers, Part E: Journal of Process Mechanical Engineering, 233*(3), 551–560. https://doi.org/10.1177/0954408918776724

Sharma, A., Jain, V., & Gupta, D. (2019b). Comparative analysis of chipping mechanics of float glass during rotary ultrasonic drilling and conventional drilling: For multi-shaped tools. *Machining Science and Technology, 23*(4), 547–568. https://doi.org/10.1080/10910344.2019.1575402

Sharma, A., Jain, V., & Gupta, D. (2019c). Tool wear analysis while creating blind holes on float glass using conventional drilling: A multi-shaped tools study. In *Advances in Manufacturing Processes: Select Proceedings of ICEMMM 2018* (pp. 175–183). Singapore: Springer. https://doi.org/10.1007/978-981-13-1724-8_17

Sharma, A., Jain, V., & Gupta, D. (2022b). Mathematical approach on chipping volume estimation generated during rotary ultrasonic drilling for float glass. *Proceedings of the National Academy of Sciences India Section A—Physical Sciences, 92*(2), 285–291. https://doi.org/10.1007/s40010-021-00732-1

Sharma, A., Jain, V., & Gupta, D. (2021). Effect of pre and post tempering on hole quality of float glass specimen: For rotary ultrasonic and conventional drilling. *Silicon, 13*, 2029–2039. https://doi.org/10.1007/s12633-020-00597-w

Sharma, A., Kalsia, M., Uppal, A. S., Babbar, A., & Dhawan, V. (2022c). Machining of hard and brittle materials: A comprehensive review. *Materials Today: Proceedings, 50*, 1048–1052. https://doi.org/10.1016/j.matpr.2021.07.452

Sharma, A., Kumar, V., Babbar, A., Dhawan, V., Kotecha, K., & Prakash, C. (2021a). Experimental investigation and optimization of electric discharge machining process parameters using grey-fuzzy-based hybrid techniques. *Materials, 14*(19), 5820. https://doi.org/10.3390/ma14195820

Sharma, A. K., Venkatesh, G., Rajesha, S., & Kumar, P. (2015). Experimental investigations into ultrasonic-assisted abrasive flow machining (UAAFM) process. *The International Journal of Advanced Manufacturing Technology, 80*(1), 477–493. https://doi.org/10.1007/s00170-015-7009-2

Sharma, V. K., Rana, M., Singh, T., Singh, A. K., & Chattopadhyay, K. (2021b). Multi-response optimization of process parameters using Desirability Function Analysis during machining of EN31 steel under different machining environments. *Materials Today: Proceedings, 44*, 3121–3126. https://doi.org/10.1016/j.matpr.2021.02.809

Shi, D., & Gibson, I. (1998). *Surface Finishing of Selective Laser Sintering Parts with Robot.* Paper Presented at the 1998 International Solid Freeform Fabrication Symposium. https://repositories.lib.utexas.edu/items/33fbcd3d-de0d-4a1d-8c1a-51117705e45a

Sidpara, A., & Jain, V. (2012). Theoretical analysis of forces in magnetorheological fluid based finishing process. *International Journal of Mechanical Sciences, 56*(1), 50–59. https://doi.org/10.1016/j.ijmecsci.2012.01.001

Singh, A., Garg, H., Kumar, P., & Lall, A. K. (2017a). Analysis and optimization of parameters in optical polishing of large diameter BK7 flat components. *Materials and*

Manufacturing Processes, *32*(5), 542–548. https://doi.org/10.1080/10426914.2016.1221103

Singh, A., Garg, H., & Lall, A. K. (2017b). Optical polishing process: Analysis and optimization using response surface methodology (RSM) for large diameter fused silica flat substrates. *Journal of Manufacturing Processes*, *30*, 439–451. https://doi.org/10.1016/j.jmapro.2017.10.017

Singh, B. P., Singh, J., Bhayana, M., Singh, K., & Singh, R. (2022). Experimental examination of the machining characteristics of Nimonic 80-A alloy on wire EDM. *Materials Today: Proceedings*, *69*, 291–296. https://doi.org/10.1016/j.matpr.2022.08.537

Singh, B. P., Singh, J., Singh, J., Bhayana, M., & Goyal, D. (2021). Experimental investigation of machining nimonic-80A alloy on wire EDM using response surface methodology. *Metal Powder Report*, *76*, S9–S17. https://doi.org/10.1016/j.mprp.2020.12.001

Singh, P., Singh, L., & Singh, S. (2020a). Analyzing process parameters for finishing of small holes using magnetically assisted abrasive flow machining process. *Journal of Bio-and Tribo-Corrosion*, *6*(1), 1–10. https://doi.org/10.1007/s40735-019-0315-8

Singh, S., Kumar, D., Ravi Sankar, M., & Jain, V. (2019). Viscoelastic medium modeling and surface roughness simulation of microholes finished by abrasive flow finishing process. *The International Journal of Advanced Manufacturing Technology*, *100*(5), 1165–1182. https://doi.org/10.1007/s00170-018-1912-2

Singh, S., Prakash, C., Pramanik, A., Basak, A., Shabadi, R., Królczyk, G., . . . Babbar, A. (2020b). Magneto-rheological fluid assisted abrasive nanofinishing of β-phase Ti-Nb-Ta-Zr Alloy: Parametric appraisal and corrosion analysis. *Materials*, *13*(22), 5156. https://doi.org/10.3390/ma13225156

Singh, S., Raj, A., Sankar, M. R., & Jain, V. (2016). Finishing force analysis and simulation of nanosurface roughness in abrasive flow finishing process using medium rheological properties. *The International Journal of Advanced Manufacturing Technology*, *85*(9), 2163–2178. https://doi.org/10.1007/s00170-015-8333-2

Singh, S., Shan, H., & Kumar, P. (2002). Wear behavior of materials in magnetically assisted abrasive flow machining. *Journal of Materials Processing Technology*, *128*(1–3), 155–161. https://doi.org/10.1016/S0924-0136(02)00442-9

Suratwala, T., Steele, R., Feit, M., Dylla-Spears, R., Desjardin, R., Mason, D., . . . Shen, N. (2014). Convergent polishing: A simple, rapid, full aperture polishing process of high quality optical flats & spheres. *JoVE (Journal of Visualized Experiments)*, *94*, e51965. https://doi.org/10.3791/51965

Tzeng, H.-J., Yan, B.-H., Hsu, R.-T., & Lin, Y.-C. (2007). Self-modulating abrasive medium and its application to abrasive flow machining for finishing micro channel surfaces. *The International Journal of Advanced Manufacturing Technology*, *32*(11), 1163–1169. https://doi.org/10.1007/s00170-006-0423-8

Umehara, N., Kirtane, T., Gerlick, R., Jain, V., & Komanduri, R. (2006). A new apparatus for finishing large size/large batch silicon nitride (Si3N4) balls for hybrid bearing applications by magnetic float polishing (MFP). *International Journal of Machine Tools and Manufacture*, *46*(2), 151–169. https://doi.org/10.1016/j.ijmachtools.2005.04.015

Vaishya, R. O., Sharma, V., Mishra, V., Gupta, A., Dhanda, M., Walia, R., . . . Burduhos-Nergis, D. P. (2022). Mathematical modeling and experimental validation of surface roughness in ball burnishing process. *Coatings*, *12*(10), 1506. https://doi.org/10.3390/coatings12101506

Vaishya, R. O., Walia, R., & Kalra, P. (2015). Design and development of hybrid electrochemical and centrifugal force assisted abrasive flow machining. *Materials Today: Proceedings*, *2*(4–5), 3327–3341. https://doi.org/10.1016/j.matpr.2015.07.158

Venkatesh, G., Sharma, A. K., & Kumar, P. (2015). On ultrasonic assisted abrasive flow finishing of bevel gears. *International Journal of Machine Tools and Manufacture*, *89*, 29–38. https://doi.org/10.1016/j.ijmachtools.2014.10.014

Wang, A., & Weng, S. (2007). Developing the polymer abrasive gels in AFM processs. *Journal of Materials Processing Technology*, *192*, 486–490. https://doi.org/10.1016/j.jmatprotec.2007.04.082

Wangikar, S. S., Patowari, P. K., Misra, R. D., & Misal, N. D. (2019). Photochemical machining: A less explored non-conventional machining process. In *Non-Conventional Machining in Modern Manufacturing Systems* (pp. 188–201). IGI Global. https://doi.org/10.4018/978-1-5225-6161-3.ch009

Xue, D., Wang, P., Jiao, L., Li, W., & Ji, Y. (2019). Experimental study on chemical mechanical polishing of chalcogenide glasses. *Applied Optics*, *58*(8), 1950–1954. https://doi.org/10.1364/AO.58.001950

Yang, H., Kannappan, S., Pandian, A. S., Jang, J.-H., Lee, Y. S., & Lu, W. (2017). Graphene supercapacitor with both high power and energy density. *Nanotechnology*, *28*(44), 445401. https://doi.org/10.1088/1361-6528/aa8948

Ye, C., Zhang, C., Zhao, J., & Dong, Y. (2021). Effects of post-processing on the surface finish, porosity, residual stresses, and fatigue performance of additive manufactured metals: A review. *Journal of Materials Engineering and Performance*, *30*(9), 6407–6425. https://doi.org/10.1007/s11665-021-06021-7

Zhong, Z.-W. (2020). Advanced polishing, grinding and finishing processes for various manufacturing applications: A review. *Materials and Manufacturing Processes*, *35*(12), 1279–1303. https://doi.org/10.1080/10426914.2020.1772481

Zhu, W.-L., & Beaucamp, A. (2020). Compliant grinding and polishing: A review. *International Journal of Machine Tools and Manufacture*, *158*, 103634. https://doi.org/10.1016/j.ijmachtools.2020.103634

Zou, Y., Xie, H., & Zhang, Y. (2020). Study on surface quality improvement of the plane magnetic abrasive finishing process. *The International Journal of Advanced Manufacturing Technology*, *109*(7), 1825–1839. https://doi.org/10.1007/s00170-020-05759-z

6 Defects during Conventional Machining of Polymer Composites

A Review

Rahul Mehra, Santosh Kumar, Satish Kumar

6.1 INTRODUCTION TO MACHINING OF POLYMER COMPOSITE MATERIALS

> In recent times, replacing metallic machine components with composite materials has been considered a viable solution to a number of problems, such as excessive metal costs, rusting, and component weight. Composite materials with physical and mechanical qualities that are comparable to or even better than those of metals are highly encouraged in the current machining industry. (Sarde and Patil, 2019; Yahaya et al., 2014; Hsissou et al., 2021)

However, polymer composite materials (PCMs) are acknowledged as a class of materials that are challenging to machine (Valino et al., 2019; Sharma et al., 2022a, 2022b). Due of differences in filler and matrix phase characteristics, PCMs have significant machining constraints (Shlykov et al., 2020; Sharma et al., 2021). As demonstrated in Figure 6.1, numerous non-traditional machining techniques have been created for processing these materials to prevent such issues (Ablyaz et al., 2021).

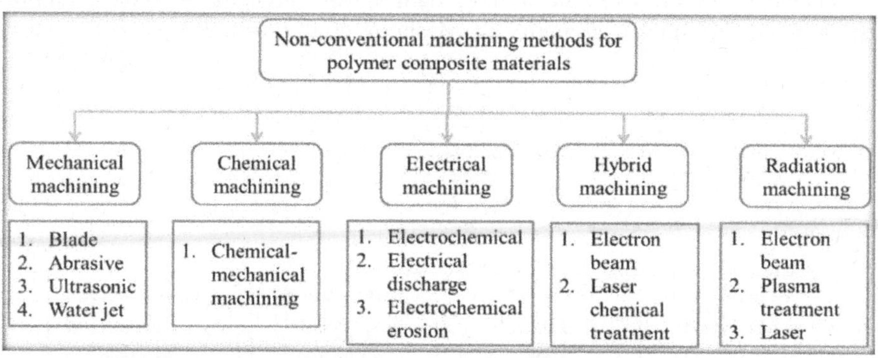

FIGURE 6.1 Classifications of non-conventional machining techniques of PCMs.

DOI: 10.1201/9781003327905-6

Although non-conventional technologies like electrochemical and chemical processing have several advantages over traditional machining techniques, they are environmentally harmful (Sharma et al., 2018a, 2019a, 2019b, 2022). "The clay bricks may be made more durable and stronger by reinforcing the clay with straw and wood, as man learned long ago. Many naturally occurring composites (wood and bones) are also available" (Ahmad, 2009). A lignin matrix holds strong cellulose fibers together in wood. Bones are made up of short fragile collagen fibers that are encased in a mineral matrix. Both wood and bones illustrate composite materials' exceptional versatility and capability in bearing loads under a variety of circumstances. So far, composite materials have been defined as materials made up of two or more elements (phases) that are macroscopically joined but not soluble in each other. In civilian, aerospace and military applications, modern synthetic composites with reinforcing fibers (one phase) and matrices (another phase) of various sorts have been developed as metal replacement materials. The invention of carbon and boron fibers in the 1960s was a watershed moment in the composites revolution (Boothroyd and Knight, 1989).

> These new fibers, which are stiffer than glass fibers, increased the rigidity of composite structures significantly. This growing adoption is due to the capacity to customise these materials to specific demands as well as their better features. Carbon-fiber-reinforced plastics were more appropriate for different applications, because of their high strength & stiffness-to-weight ratio. Glass fibers' better resilience to environmental assault makes glass fiber reinforced polymers more appealing for maritime applications, as well as the chemical and food sectors. (Chawla, 1987; Mathews and Rawlings, 1994)

As a result, as indicated in Figure 6.2, the current review is separated into distinct sections.

6.2 LITERATURE SURVEY

Conventional machining of fiber composites like drilling, milling, turning, etc., has been reviewed. Various researchers have investigated machining using different mechanical properties like flexural strength, tensile strength, inter shear laminar strength (ILSS), impact strength, etc. Meshram et al. (2018) compared the tensile

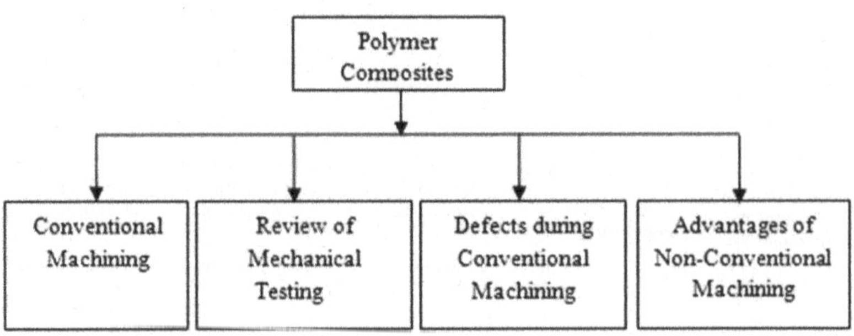

FIGURE 6.2 Classification of composites review area.

strength and drilling thrust force between nylon reinforced composite and pure epoxy (Figure 6.3).

Figure 6.3 shows tensile testing results. They concluded that the tensile strength of nylon reinforced composites has enhanced mainly due to the nylon mat presence.

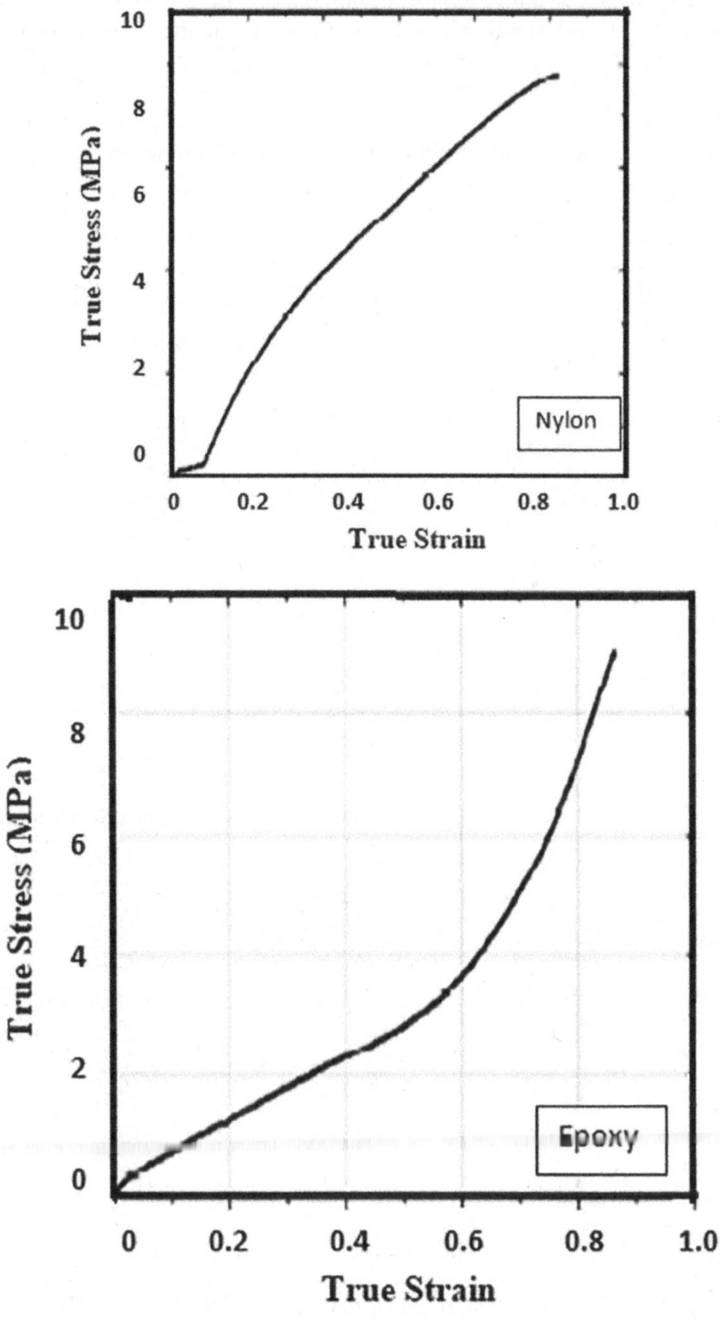

FIGURE 6.3 Tensile testing of nylon and pure epoxy.

Moreover, there is not much variation in the values of thrust forces between nylon reinforced and pure epoxy composites.

Thereafter, Nayak et al. (2014) carried out flexural testing on TiO2-, SiO2- and Al2O3-reinforced glass fiber hybrid composites. It is quite clear from Figure 6.4 that flexural strength and flexural modulus of SiO2-based epoxy composites are more as compared to Al2O3 and titania-based composites. It is mainly due to the finer particle size of silica (Nayak et al., 2014).

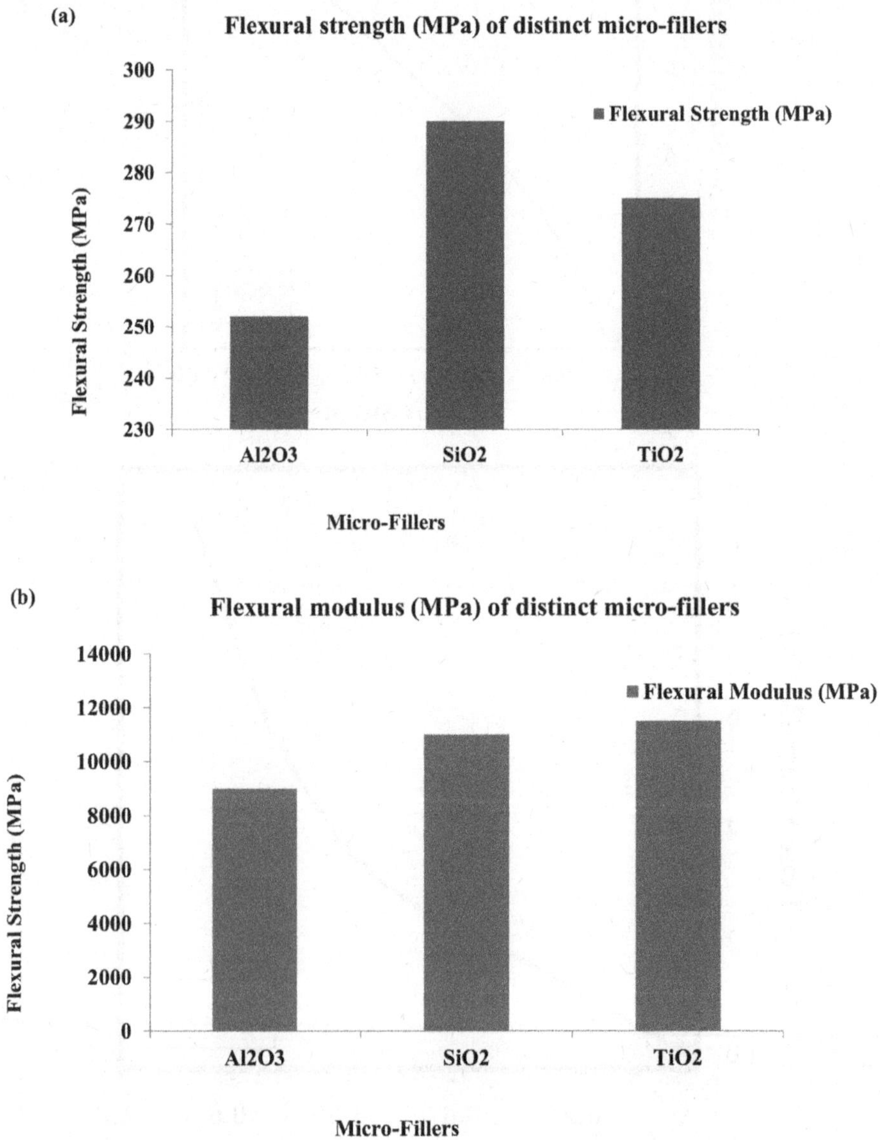

FIGURE 6.4 Flexural testing of glass fiber using epoxy modifiers.

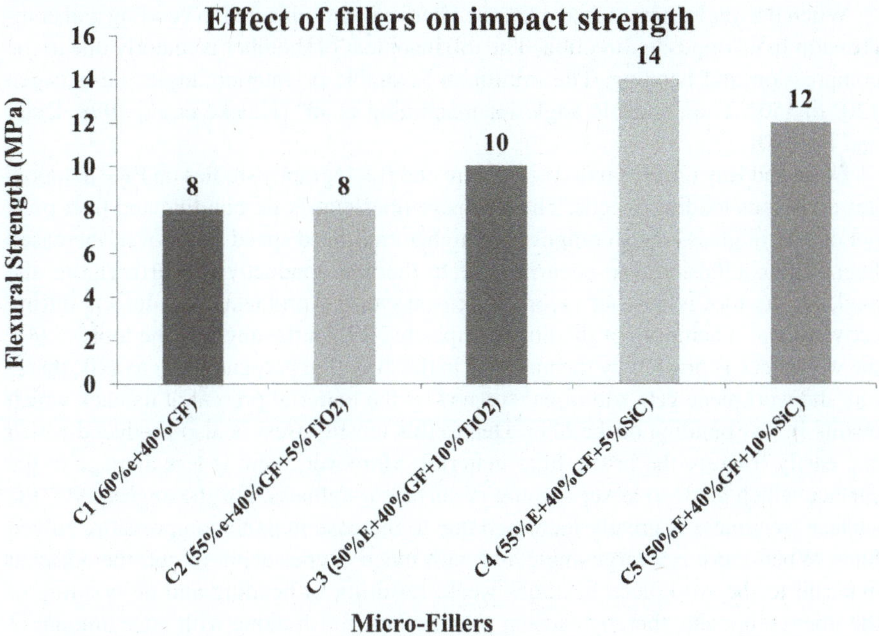

FIGURE 6.5 Impact testing of various glass fibers.

Moreover, Suresh et al. (2017) measured impact strength of various composites with varying weight percentage of epoxy, glass fiber, epoxy, glass fiber, titanium and silicon powder (Figure 6.5).

It is quite clear from Figure 2.3 that 5% of silicon carbide composite have maximum impact strength.

6.3 DEFECTS DURING CONVENTIONAL MACHINING OF COMPOSITES

The various problems arise during conventional machining of composites include rapid tool wear, delamination, pull out of fiber, deformation and deterioration of the machined surface (Abrate, 1997; Boldt and Chanani, 1987).

These machining problems also occur due to the fiber orientation of reinforced composite with the matrix. Fiber placed in perpendicular direction results in large number of fractures and chips, whereas in parallel orientation, good machining with smooth surface is observed with small amount of cracks in it (Koplev et al., 1983). The composite materials are basically non-homogeneous and mostly anisotropic. The development of machining force and stresses during machining is majorly influenced by the angle of fiber orientation.

The forces increase when the angle of fiber orientation is in between 0 to 60°, decreases when it lies between 60 to 120° and again increases largely above 120°.

When the angle increases to 135°, the fiber is again subjected to bending and compression in its opposite direction. The infringement of the fiber is majorly due to the compression and bending. The minimum favorable orientation angles lie between 120° to 150°. The amicable angle for machining is 45° (Klocke et al., 1998; Dalai and Ray, 2011).

Dalai and Ray (2011) carried out failure and fractography studies on FRP composites at various loading speeds. The extensive fiber/matrix de-bonding and fiber pull-out occurs in glass/epoxy composite at higher crosshead speed. Moreover, increased fiber/matrix adhesion also occurred due to thermal conductivity. Furthermore, the angle of the tool is another favorable reason causing problems and defects during conventional machining of the fiber composites. When the angle of the tool is <90°, the workpiece is pressed by the machine in the direction perpendicular to axis, therefore the workpiece gets additional support by the material present at its back which results in less bending of the fiber. Due to this tensile stress is also produced which can easily fracture the brittle fiber material. Moreover, there is less damage to the surface which leads to lesser amount of surface roughness. For the angles >90°, the surface roughness is greatly increased due to increase in axial compression, bulged fiber. When there is a large angle and with much deeper depth of cut, the adjacent material to the workpiece becomes weak, resulting in bending and de-bonding of the fiber composite, thereby causing poor surface finish along with huge amount of damage to the inner surface of the fiber (Wang et al., 2011).

Various researchers have focused on turning of hard to cut materials. In this context, Ramulu (1998) investigated that high tool wear rate (TWR) and poor surface of FRPs occurred during turning with HSS and carbide as the tool materials. This occurs mainly due to abrasive and inhomogeneous nature of material. Turning with ceramic tooling is not recommended because of its low thermal conductivity. PCD (polycrystalline diamond) has been used in turning of FRPs and shows satisfactory but limited results due to its better tool geometry as compared to other turning tools. Diamond abrasive cutters were also used during turning of extremely difficult to machine GFRPs due to its tremendous abrasive nature (Faridnia et al., 1989).

Furthermore, turning operation was carried out on three different types of reinforced fiber composites (glass, carbon and aramid) using K20 tool made of carbide. It has been observed that reinforcements in the fiber were prominent reason for the cracks or rupture, deformation, delamination and shearing during the material removal mechanism (Srivatsan and Bowden, 1992).

Moreover, variation in temperature is noticed during turning of fiber reinforcement composite (Sreejith et al., 1998). At 100 mm/min turning speed, minimum rise in temperature has been observed, whereas, at higher turning speeds, greater rise in temperature has been observed.

Various authors have also focused on milling operation of FRPs. It has been observed that during milling operation, low volume ratio is removed as compared to conventional drilling. Due to this reason, milling operation is considered more fit than the drilling for high-speed machining and low feed rates (Ali et al., 2013). Also, the quality of machining surface is deeply affected by the abrasive nature and friction between the workpiece and the tool. Machining parameters combinations like low feed and high speed even leads to the chipping and melting of matrix or can even burn. The thermal stress produced on the tools during the machining is a serious problem while using small diameter tools. FRPs are low thermal conductivity materials and have to absorb large amounts of heat generated during the conventional milling operation. Therefore, if the ability of the tool to absorb the generated heat is low, there is increase in large amount of friction, resulting in high thermal stresses, greater tool wear along with unsafe operational conditions.

During grinding of reinforced fibers, huge amount of problems like fiber pullout, delamination, flaming and rough projections like burrs and burning is observed. With deeper depth of grinding, more damage is done to the material surface. Therefore, grinding operation leads to large problems as compared to other conventional machining processes (Hu and Zhang, 2001). Kim and Lee (2005) uses dry and wet grinding conditions for grinding the reinforced composites. During the experimentation, the direction of fiber orientation was kept parallel and perpendicular. The rise in temperature was observed more up to 280°C in dry condition, which causes matrix degradation. Therefore, to avoid this wet condition is more preferred. The temperature during the wet condition reaches minimum up to 60°C, which in turn helps to avoid the degradation and higher surface roughness. Therefore, grinding process is more preferred for machining of reinforced materials and possesses high resistance to heat (Tonshoff et al., 1998).

Various researchers have focused on drilling of FRPs. Due to in-homogeneous and anisotropical nature of fiber reinforced material, drilling also raises particular issues related to damages like peel up at entrance, peel out at exit, inter laminar cracking, matrix de-bonding, etc. Due to these problems and defects, the tool wear rate is very high which can affect the production rate (Persson et al., 1997; Hamdoun et al., 2004).

Delamination occurs between the adjacent layers in the intra laminar parts of the material. The defect not only depends upon the character of the fiber material but also upon the epoxy and its properties. Mechanism of delamination is characterized into two types, viz. peel-up delamination mechanism at entrance and push-out delamination at exit. Mechanism of peel up is caused by drilling up thrust forces which pushes the abraded material to the surface. In the initial phase, the cutting edge will erode the laminate, and as the drill starts to move deeper in the surface, the material will start spiraling up before the actual drilling takes place. This will create an upward peeling force which will split the plate of the laminas. On the other hand, push-out mechanism is the drilling resultant of compressive thrust force which always exerts downwards pressure. The drill bit after crossing the upper plies breaks the inter laminar bond in the region of the hole and tends to approach at the other end of the material surface. At this point, the downwards thrust force of the drill overpowers the strength of fiber bonding and completely penetrates into the other side of the composite material (Hocheng and Dharan, 1990).

As a result, there are various types of drilling like pilot hole drilling, backup drilling in which special drill bits made of polycrystalline diamond (PCD) and carbide are used to avoid delamination and tool wear which generally happens due to the high abrasiveness of the composite fiber. But this also leads to increase in the overall cost of the drilling and makes the production process more expensive (Taylor, 2000; Schulze et al., 2011).

Another severe issue with machining of FRC is the formation of dust leading to rigorous health problems for both humans as well as for machines and requiring extra caution while extracting and filtrating (Tandon et al., 1990; Lemma et al., 2002).

The thermal stresses developed while drilling also assist in the delamination defect by softening of the matrix material. De-bonding, tearing of fiber and inter laminar cracks of the are also observed along the side walls fiber and matrix. Composites due to less thermal conductivity and high abrasive nature cause heating of the tool, resulting in heavy tool wear. While drilling epoxy, almost 50% of thermal energy is absorbed by the tool and the rest is uniformly absorbed by workpiece and chips formation during the machining. While drilling metals, 75% is eliminated with chips formation, 18% by tool and the rest 7% by workpiece. Circularity of the hole should be observed after the drilling process is completed because it has the possibility of the drilled hole to cause the deformation of the hole. It has the tendency to again return to its original position, causing tightening of the hole around the drilling area. Anisotropy is the major cause for this defect in the composite material while conventional drilling. These aforementioned problems and defects are very much accountable for the rejection of the part production which leads to increase in the overall cost of the product (Piquet et al., 2000).

Moreover, the grinding variables also have an effect on the grinding chips in grinding of two-dimensional carbon/silicon carbide composite (Liu et al., 2017).

Furthermore, various researchers performed drilling of GFRPs with HSS drills. In this context, Tagliaferri et al. (1990) analyzed that damage is done due to the variation in feed and cutting speed. Thereafter, a number of GFRP plates were drilled by Bongiorno et al. (1998) using 5 mm diameter HSS drills. Results revealed that various cracks along the machined holes were formed that greatly affects the fatigue property of the material. Therefore, cracking can be avoided by using low feed rate and support plates at the back, resulting in reduction in push-out defects. An experimental study on GFRPs has been done to examine the outcome of process parameters like speed and feed on thrust force and delamination (Khashaba, 1996). The result shows that the process of push-out delamination is more rigorous than the peel up. Also, the size of delamination area increases with the rise in feed and decreasing speed. The change in matrix material and orientation causes great effect on the delamination (Sonbaty et al., 2004; Velayudham et al., 2005). The thrust forces and torque were used as the processing parameters. The result shows that huge amount of delamination occurred at higher torque. During drilling with carbide tools, surface roughness (SR), thrust force and other damages were also analyzed by Davim et al. (2004). Results show that the SR rises with the increase of feed rate and diminishes in the cutting speed.

The machining of hybrid GFRPs is difficult due to their non-homogeneous structure and harder reinforcements. Conventional methods like milling, drilling and turning are not able to machine these types of advanced materials (Mathews and Rawling, 1999; Nayak et al., 2014; Bhoopathi et al., 2014). Various problems like tool wear, low surface quality, de-lamination, pulling out of fiber and matrix recession may occur during machining of these types of materials. This is also mainly due to high disassociation temperature and high thermal conductivity (Muller and Monaghan, 2000; Birhan and Ekici, 2010; Liu et al., 2012; Kashwani and Al-Tamimi, 2014).

6.4 NON-CONVENTIONAL MACHINING OF POLYMER COMPOSITES

The carbon is frequently combined with a matrix such as resin, metal or ceramic to make a structural composite due to its appealing qualities, which include low-density and high-density modulus and strength, low thermal expansion coefficient, high thermal conductivity, etc. The use of fiber-reinforced composites is currently widespread in numerous industries. The schematics illustration of FRCs (Du et al., 2019) is shown in Figure 6.6.

The CFRCs is widely employed in aircraft, advanced structural, energy, break system, defense applications, etc., owing to its low weight, corrosion resistance, chemical stability, wear resistance, specific strength, etc. (Du et al., 2019).

Problems and defects as discussed in the previous literature related to conventional machining can be minimized to a great extent by using non-conventional machining methods. It is very much necessary to understand the various alternative non-conventional machining methods to achieve better and optimized results, including good surface finish.

There are various reasons like abrasiveness, poor machinability, high hardness, cracks, stresses and lower thermal resistance which arise the need to use the advanced machining techniques.

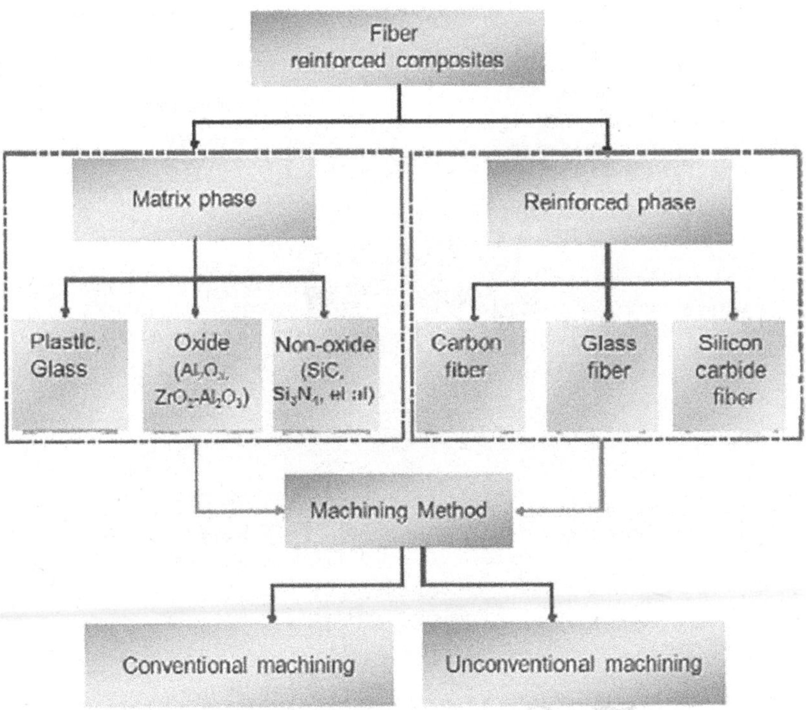

FIGURE 6.6 The schematic illustrations of FRCs.

To obtain complex shape, high precision accuracy in small or micro parts and better surface finish, non-conventional methods are in much demand. The various alternative non-conventional methods which can be used are ultrasonic machining (USM), chemical machining (CM), etc.

The development of micro features on polymer materials can be done using a variety of manufacturing processes (Rawal et al., 2022) as shown in Figure 6.7.

Many of these procedures call for additional steps to render the final result useful. The total cost of production goes higher as a result of some procedures requiring expensive initial setup fees and specialized environments like vacuum chambers and clean rooms. Due to material ablation, laser processing results in uneven polymer composition and non-uniform micro channel surface. Hazardous chemicals are used in the chemical process. As a result, traditional micro machining techniques like micro drilling/milling can efficiently and quickly produce complex shapes (Malayath et al., 2019).

Minimum chip thickness, workpiece microstructure, tool edge radius, specific cutting force, surface quality and dynamic instability are a few of the crucial variables [64]. Surface finish, machinability and material removal rate are three metrics that can be used to assess how well the machining performed. In the fish bone

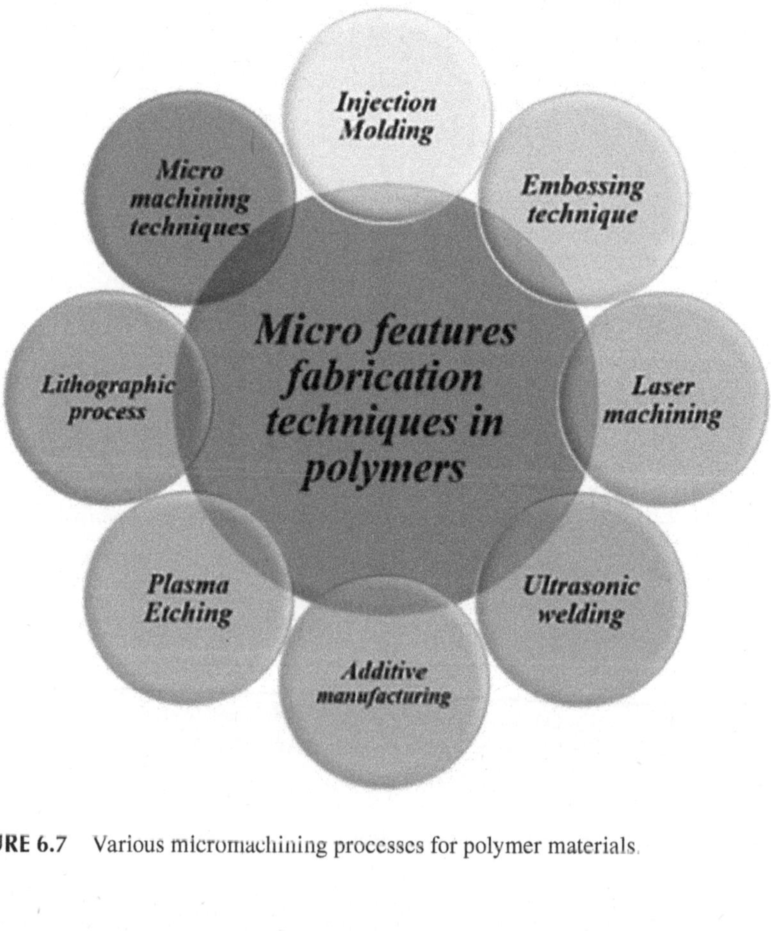

FIGURE 6.7 Various micromachining processes for polymer materials.

diagram (Figure 6.8), various process variables that affect how polymer composites are machined are depicted (Rawal et al., 2022).

The different properties of the different materials decide the applicability of using that advance machining process. All conventional processes are not able to machine the polymer composites owing to the presence of hard reinforcements. As a result, it is very important to understand the machining nature and purpose of both the materials and machine. Therefore, non-conventional machining methods has been adopted by various small- and large-scale industries due to its various advance features like noncontact of tool with the workpiece, can machine the hard composite and fragile parts like glass, use of simple clamping devices, high-energy density of up to 106 W/mm^2, etc. This causes vaporization of the material at the localized areas which helps in reduction of the heat-affected zone. Figure 6.9 shows the cause-and-effect diagram for laser beam machining to achieve higher quality products.

The thermal non-conventional method like laser-based machining has been considered the best alternative for machining the hard to cut materials (Maclean et al., 2018; Joshi and Sharma, 2018). Various authors have also used thermal and physical surface modification techniques such as thermal spray coatings. These also enhanced the erosion, corrosion and wear resistance of the machined surface (Bedi et al., 2019; Kumar and Kumar, (2018); Kumar et al., 2019a, 2019 b, 2020a ,2021).

This study provides a comprehensive view of the composite materials and their machining processes, whereas, some other trending machining processes and additive manufacturing processes provide details about the defects and damages during these processes. Several machinability criteria (i.e. cutting forces, tool wear, delamination and surface finish) were reported for getting required machined substrates (Akhai and Rana, 2022; Babbar et al., 2020a, 2020b, 2020c, 2020d, 2021, 2022; Kalia et al., 2022; Khanduja et al., 2021; Kumar et al., 2021b, 2021c; Prakash et al., 2021; Rampal et al., 2021; Rana and Akhai, 2022; Sharma et al., 2018b, 2019c,

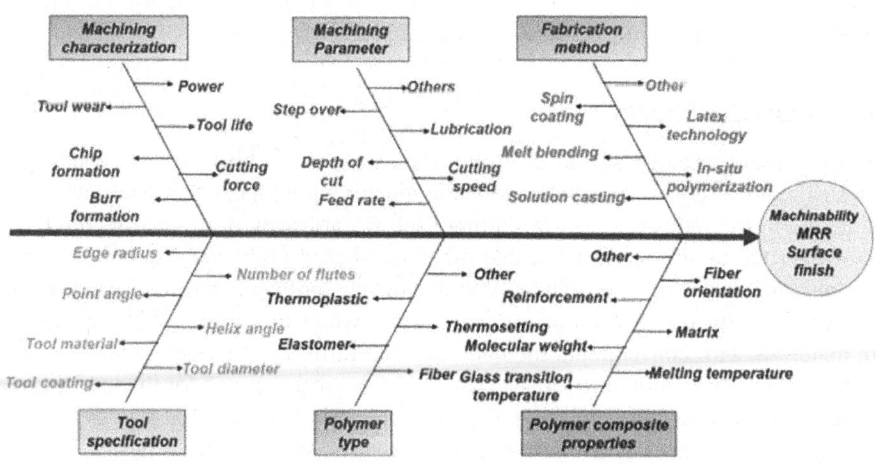

FIGURE 6.8 Fish bone chart illustrating various factors that affects the machining of polymer composite.

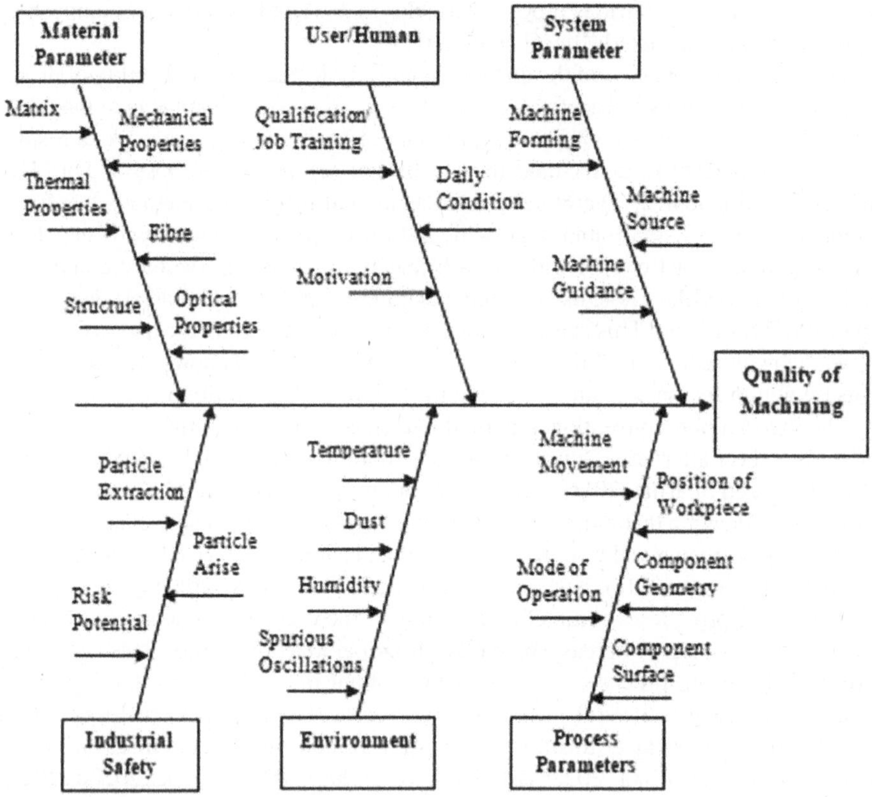

FIGURE 6.9 Cause-effect diagram of laser beam machining.

2020a, 2020b, 2021a, 2021b, 2022c, 2022d, 2023a, 2023b, 2023c; Sharma and Jain, 2020; Sharma, V.K. et al., 2021; Singh et al., 2021, 2022, 2023).

6.5 CONCLUSIONS

In this present research scenario, machining of advanced fiber-reinforced materials still requires a very big challenge while machining with conventional methods. In addition, the industrial cost of cutting such materials by traditional methods is also high due to shorter tool life and other tool-based problems. Therefore, from the outcome, it has been concluded that the various problems like tool wear, fiber pullout, fiber fracture, delamination, etc., arises while machining of the composite materials with conventional methods can be minimized to much greater extent by using non-conventional processes, which are in great demand for the machining of composites. The non-conventional process majorly uses thermal energy as the source of heat for melting or vaporizing the surface material to get the desired result. It has been observed that various research works have been performed by using non-conventional machining which shows great results, including high material removal rate, low surface roughness along with fine precision. Moreover, the combination of various process parameters to machine hybrid polymers using non-conventional machining provides all the necessary outcomes which helps in getting the desired values at its best which has not been reported till yet by conventional methods.

REFERENCES

Ablyaz, T.R., Shlykov, E.S., Muratov, K.R., & Sidhu, S.S. (2021). Analysis of Wire-Cut Electro Discharge Machining of Polymer Composite Materials. *Micromachines*, 12, 1–14. https://doi.org/10.3390/mi12050571

Abrate, S. (1997). Machining of Composite Materials. In *Composites Engineering Handbook* (pp. 777–809). New York: CRC Press.

Ahmad, J. (2009). *Machining of Polymer Composites*. Springer. https://doi.org/10.1007/978-0-387-68619-6

Akhai, S., & Rana, M. (2022). Taguchi-Based Grey Relational Analysis of Abrasive Water Jet Machining of Al-6061. *Materials Today: Proceedings*, 65, 3165–3169. https://doi.org/10.1016/j.matpr.2022.05.361

Ali, H.M., Iqbal, A., & Liang, L.I. (2013). A Comparative Study on the Use of Drilling and Milling Processes in Hole Making of GFRP Composite. *Sadhana*, 38, 743–760.

Babbar, A., Jain, V., Gupta, D., Prakash, C., & Sharma, A. (2020b). Fabrication and Machining Methods of Composites for Aerospace Applications. In *Characterization, Testing, Measurement, and Metrology* (1st ed., pp. 109–124). Boca Raton, FL: CRC Press.

Babbar, A., Jain, V., Gupta, D., Prakash, C., Singh, S., & Sharma, A. (2020a). 3D Bioprinting in Pharmaceuticals, Medicine, and Tissue Engineering Applications. In *Advanced Manufacturing and Processing Technology* (1st ed., pp. 147–161). Boca Raton, FL: CRC Press, 2021.

Babbar, A., Jain, V., Gupta, D., Prakash, C., Singh, S., & Sharma, A. (2020c). Effect of Process Parameters on Cutting Forces and Osteonecrosis for Orthopedic Bone Drilling Applications. In *Characterization, Testing, Measurement, and Metrology* (1st ed., pp. 93–108). Boca Raton, FL: CRC Press.

Babbar, A., Jain, V., Gupta, D., & Sharma, A. (2020d). Fabrication of Microchannels Using Conventional and Hybrid Machining Processes. In *Non-Conventional Hybrid Machining Processes* (1st ed., pp. 37–51). Boca Raton, FL: CRC Press.

Babbar, A., Rai, A., & Sharma, A. (2021). Latest Trend in Building Construction: Three-Dimensional Printing. *Journal of Physics: Conference Series*, 1950, 012007.

Babbar, A., Sharma, A., & Singh, P. (2022). Multi-Objective Optimization of Magnetic Abrasive Finishing Using Grey Relational Analysis. *Materials Today: Proceedings*, 50, 570–575.

Bedi, T.S., Kumar, S., & Kumar, R. (2019). Corrosion Performance of Hydroxyapatite and Hydroxyapaite/Titania Bond Coating for Biomedical Applications. *Materials Research Express*, 7, 015402. https://doi.org/10.1088/2053-1591/ab5cc5

Bhoopathi, R., Ramesh, M., & Deepa, C. (2014). Fabrication and Property Evaluation of Banana-Hemp-Glass Fiber Reinforced Composites. *Procedia Engineering*, 97, 2032–2041.

Birhan, I., & Ekici, E. (2010). Experimental Investigations of Damage Analysis in Drilling of Woven Glass Fiber-Reinforced Plastic Composites. *International Journal of Advanced Manufacturing Technology*, 49(9), 861–869.

Boldt, J.A., & Chanani, J.P. (1987). Solid-Tool Machining and Drilling. In *Engineered Materials Handbook* (Vol. 1, pp. 667–672). Ohio 44104, United States: ASM International. Handbook Committee.

Bongiorno, A., Capello, E., Copani, G., & Tagliaferri, V. (1998). Drilled Hole Damage and Residual Fatigue Behaviour of GFRP. *Proceedings of the ECCM-8 Conference* (Vol. 2, pp. 525–532). https://art.torvergata.it/handle/2108/111651

Boothroyd, G., & Knight, W. (1989). *Fundamentals of Machining and Machine Tools* (2nd ed.). New York: Marcel Dekker.

Chawla, K.K. (1987). *Composite Materials: Science and Engineering*. New York: Springer.

Dalai, R.P., & Ray, B.C. (2011). Failure and Fractography Studies of FRP Composites: Effects of Loading Speed and Environments. *Processing and Fabrication of Advanced Materials*, 19, 1–9.

Davim, J.P., Reis, P., & António, C.C. (2004). Experimental Study of Drilling Fiber Reinforced Plastics (GFRP) Manufactured by Hand Lay-Up. *Composites Science and Technology*, 64, 289–297.

Du, J., Zhang, H., Geng, Y., Ming, W., He, W., Ma, J., Cao, Y., Li, X., & Liu, K. (2019). A Review on Machining of Carbon Fiber Reinforced Ceramic Matrix Composites. *Ceramics International*. https://doi.org/10.1016/j.ceramint.2019.06.112

Faridnia, M., Garbini, J.L., & Jorgensen, J.E. (1989). Machining of Graphite/Epoxy Materials with Polycrystalline Diamond Tools. *Machining Characteristics of Advanced Materials*, 33–39.

Hamdoun, Z., Guillaumat, L., & Lataillade, J.L. (2004). Influence of the Drilling on the Fatigue Behaviour of Carbon Epoxy Laminates. *ECCM 11*, Rhodes, Greece.

Hocheng, H., & Dharan, C.K.H. (1990). Delamination During Drilling in Composite Laminates. *Journal of Engineering for Industry*, 112, 236–239.

Hsissou, R., Benhiba, F., Echihi, S., Benkhaya, S., Hilali, M., & Berisha, A. (2021). New Epoxy Composite Polymers as a Potential Anticorrosive Coating for Carbon Steel in 3.5% NaCl Solution: Experimental and Computational Approaches. *Chemical Data Collections*, 31, 100619.

Hu, N., & Zhang, L. (2001). Grindability of Unidirectional Carbon Fiber-Reinforced Plastics. *ICCM-13*, Beijing, China.

Joshi, P., & Sharma, A. (2018). Optimization of Dimensional Accuracy for the Nd YAG Laser Cutting of Aluminium Alloy Thin Sheet Using a Hybrid Approach. *Lasers in Engineering*, 41(8), 263–281.

Kalia, G., Sharma, A., & Babbar, A. (2022). Use of Three-Dimensional Printing Techniques for Developing Biodegradable Applications: A Review Investigation. *Materials Today: Proceedings*, 62, 346–352.

Kashwani, G.A., & Al-Tamimi, A.K. (2014). Evaluation of FRP Bars Performance Under High Temperature. *Physics Procedia*, 55, 296–300.

Khanduja, P., Bhargave, H., Babbar, A., Pundir, P., & Sharma, A. (2021). Development of Two-Dimensional Plotter Using Programmable Logic Controller and Human Machine Interface. *Journal of Physics: Conference Series*, 1950, 012012.

Khashaba, U.A. (1996). Notched and Pin Bearing Strength of GFRP Composite Laminates. *Journal of Composite Materials*, 30, 2042–2055.

Kim, G., & Lee, K.Y. (2005). Critical Thrust Force at Propagation of Delamination Zone Due to Drilling of FRP/Metallic Strips. *Composite Structures*, 69, 137–141.

Klocke, F., Koenig, W., Rummenhoeller, S., & Wuertz, C. (1998). *Milling of Advanced Composites, Machining of Ceramics and Composites* (Ed. Marcel Dekker, pp. 249–266). New York: Marcel Dekker.

Koplev, A., Lystrup, A., & Vorm, T. (1983). The Cutting Process, Chips, and Cutting Forces in Machining CFRP. *Composites*, 14, 371–376.

Kumar, M., Kant, S., & Kumar, S. (2019a). Corrosion Behavior of Wire Arc Sprayed Ni-based Coatings in Extreme Environment. *Materials Research Express*, 6, 106427. https://doi.org/10.1088/2053-1591/ab3bd8

Kumar, R., & Kumar, S. (2018). Comparative Parabolic Rate Constant and Coating Properties of Nickel, Cobalt, Iron and Metal Oxide Based Coating: A Review. *I-Manager's Journal on Material Science*, 6(1), 45–56. https://doi.org/10.26634/jms.6.1.14379

Kumar, S., Handa, A., Chawla, V., Grover, N.K., & Kumar, R. (2021c). Performance of Thermal- Sprayed Coatings to Combat Hot Corrosion of Coal-Fired Boiler Tube and Effects of Process Parameters and Post Coating Heat Treatment on Coating Performance: A Review. *Surface Engineering*, 37(5), 1–28.

Kumar, S., Kumar, M., & Handa, A. (2018). Combating Hot Corrosion of Boiler Tubes: A Study. *Journal of Engineering Failure Analysis*, 94, 379–395.

Kumar, S., Kumar, M., & Handa, A. (2019a). Comparative Study of High Temperature Oxidation Behavior of Wire Arc Sprayed Ni-Cr and Ni-Al Coatings. *Engineering Failure Analysis*, 106, 104173–104189.

Kumar, S., Kumar, M., & Handa, A. (2019b). High Temperature Oxidation and Erosion-Corrosion Behaviour of Wire Arc Sprayed Ni-Cr Coating on Boiler Steel. *Material Research Express*, 6, 125533.

Kumar, S., Kumar, M., & Handa, A. (2020a). Erosion Corrosion Behavior and Mechanical Property of Wire Arc Sprayed Ni-Cr and Ni-Al Coating on Boiler Steels in Actual Boiler Environment. *Material at High Temperature*, 37(6), 1–15 https://doi.org/10.1080/09603409.2020.1810922

Kumar, S., Sudhakar, R.P., Goyal, D., & Sehgal, S. (2021b). Process Modelling for Machining Inconel 825 Using Cryogenically Treated Carbide Insert. *Metal Powder Report*, 76, 66–74. https://doi.org/10.1016/j.mprp.2020.06.001

Lemma, E.L., Chen, E.S., & Wang, J. (2002). Study of Cutting Fiber-Reinforced Composites by Using Abrasive Water-Jet with Cutting Head Oscillation. *Composite Structures*, 57(1), 297–303.

Liu, D.F., Tang, Y.J., & Cong, W.L. (2012). A Review of Mechanical Drilling for Composite Laminates. *Composite Structure*, 94(4), 1265–1279.

Liu, Q., Huang, G.Q., Xu, X.P., Fang, C.F., & Cui, C.C. (2017). A Study on the Surface Grinding of 2D C/SiC Composites. *The International Journal of Advanced Manufacturing Technology*, 93(1), 1–9.

Maclean, J.O., Hodson, J.R., Tangkijcharoenchai, C., Al-Ojaili, S., Rodsavas, S., Coomber, S., & Voisey, K.T. (2018). Laser Drilling of Microholes in Single Crystal Silicon Using Continuous Wave (CW) 1070 nm Fiber Lasers with Millisecond Pulse Widths. *Lasers in Engineering*, 39(1–2), 53–65.

Malayath, G., Narayanan, J.K., Sidpara, A.M., & Deb, S. (2019). Experimental and Theoretical Investigation into Simultaneous Deburring of Microchannel and Cleaning of the Cutting Tool in Micromilling. *Proceedings of the Institution of Mechanical Engineers, Part B: Journal of Engineering Manufacture*, 233, 1761–1771. https://doi.org/10.1177/0954405418798864

Mathews, F.L., & Rawlings, R.D. (1994). *Composite Materials: Engineering and Science*. London: Chapman & Hall.

Mathews, F.L., & Rawlings, R.D. (1999). *Composite Materials* (1st ed.). Engineering and Science. London: Woodhead Publishing.

Meshram, P., Sahu, S., Ansari, M.Z., & Mukherjee, S. (2018). Study on Mechanical Properties of Epoxy and Nylon/Epoxy Composite. *(ICMPC 2017): Materials Today Proceedings*, 5(2), 5925–5932.

Muller, F., & Monaghan, J. (2000). Non-Conventional Machining of Particle Reinforced Metal Matrix Composite. *International Journal of Machine Tool and Manufacture*, 40(9), 1351–1366.

Nayak, R.K., Dasha, A., & Ray, B.C. (2014). Effect of Epoxy Modifiers (Al2O3/SiO2/TiO2) on Mechanical Performance of Epoxy/Glass Fiber Hybrid Composites. Proc. 3rd International Conference on Materials Processing and Characterization (ICMPC). *Procedia Materials Science*, 6, 1359–1364.

Persson, E., Eriksson, I., & Zackrisson, L. (1997). Effects of Hole Machining Defects on Strength and Fatigue Life of Composite Laminates. *Composites A*, 28, 141–151.

Piquet, R., Ferret, B., Lachaud, F., & Swider, P. (2000). Experimental Analysis of Drilling Damage in Thin Carbon/Epoxy Plate Using Special Drills. *Composites A*, 31, 1107–1115.

Prakash, C., Kumar, V., Mistri, A., Sharma, A., Uppal, A.S., Babbar, A., & Pathri, B.P. (2021). Investigation of Functionally Graded Adherents on Failure of Socket Joint of FRP Composite Tubes. *Materials*, 14, 6365.

Rampal, R., Goyal, T., Goyal, D., Mittal, M., Dang, R.K., & Bahl, S. (2021). Magneto-Rheological Abrasive Finishing (MAF) of Soft Material Using Abrasives. *Materials Today: Proceedings*, 45, 51140–5121. https://doi.org/10.1016/j.matpr.2021.01.629

Ramulu, M. (1998). Cutting-Edge Wear of Polycrystalline Diamond Inserts in Machining of Fibrous Composite Material. In *Machining of Ceramics and Composites* (pp. 357–410). New York: Marcel Dekker.

Rana, M., & Akhai, S. (2022). Multi-Objective Optimization of Abrasive Water Jet Machining Parameters for Inconel 625 Alloy Using TGRA. *Materials Today: Proceedings*, 65, 3205–3210. https://doi.org/10.1016/j.matpr.2022.05.374

Rawal, S., Sidpara, A.M., & Paul, J. (2022). A Review on Micro Machining of Polymer Composites. *Journal of Manufacturing Processes*, 77, 87–113. https://doi.org/10.1016/j.jmapro.2022.03.014

Sarde, B., & Patil, Y.D. (2019). Recent Research Status on Polymer Composite Used in Concrete—An Overview. *Materials Today: Proceedings*, 18, 3780–3790.

Schulze, V., Becke, C.K., & Dietrich, S. (2011). Machining Strategies for Hole Making in Composites with Minimal Workpiece Damage by Directing the Process Forces Inwards. *Journal of Materials Processing Technology*, 211(3), 329–338.

Sharma, A., Babbar, A., Jain, V., & Gupta, D. (2018b). Enhancement of Surface Roughness for Brittle Material During Rotary Ultrasonic Machining. *MATEC Web of Conferences*, 249, 01006.

Sharma, A., Babbar, A., Tian, Y., Pathri, B.P., Gupta, M., & Singh, R. (2022b). Machining of Ceramic Materials: A State of the Art Review. *International Journal on Interactive Design and Manufacturing (IJIDeM)*, 1–21.

Sharma, A., Fidan, I., Huseynov, O., Ali, M.A., Alkunte, S., Rajeshirke, M., Gupta, A., Hasanov, S., Tantawi, K., Yasa, E., Yilmaz, O., Loy, J., & Popov, V. (2023b). Recent Inventions in Additive Manufacturing: Holistic Review. *Inventions*, 8(4), 103. https://doi.org/10.3390/inventions8040103

Sharma, A., Grover, V., Babbar, A., & Rani, R. (2020b). A Trending Nonconventional Hybrid Finishing/Machining Process. In *Non-Conventional Hybrid Machining Processes* (1st ed., pp. 79–93). Boca Raton, FL: CRC Press.

Sharma, A., & Jain, V. (2020). Experimental Investigation of Cutting Temperature During Drilling of Float Glass Specimen. In *IOP Conference Series: Materials Science and Engineering* (Vol. 715, No. 1, p. 012050). IOP Publishing. https://doi.org/10.1088/1757-899X/715/1/012050

Sharma, A., Jain, V., & Gupta, D. (2018a). Characterization of Chipping and Tool Wear During Drilling of Float Glass Using Rotary Ultrasonic Machining. *Measurement*, 254–263.

Sharma, A., Jain, V., & Gupta, D. (2019a). Comparative Analysis of Chipping Mechanics of Float Glass During Rotary Ultrasonic Drilling and Conventional Drilling: For Multi-Shaped Tools. *Machining Science and Technology*, 23, 547–568.

Sharma, A., Jain, V., & Gupta, D. (2019b). Multi-Shaped Tool Wear Study During Rotary Ultrasonic Drilling and Conventional Drilling for Amorphous Solid. *Proceedings of the Institution of Mechanical Engineers, Part E: Journal of Process Mechanical Engineering*, 233, 551–560.

Sharma, A., Jain, V., & Gupta, D. (2019c). Tool Wear Analysis While Creating Blind Holes on Float Glass Using Conventional Drilling: A Multi-Shaped Tools Study. In *Advances in Manufacturing Processes: Select Proceedings of ICEMMM 2018* (pp. 175–183). Singapore: Springer. https://doi.org/10.1007/978-981-13-1724-8_17

Sharma, A., Jain, V., & Gupta, D. (2021a). Effect of Pre and Post Tempering on Hole Quality of Float Glass Specimen: For Rotary Ultrasonic and Conventional Drilling. *Silicon*, 13, 2029–2039.

Sharma, A., Jain, V., & Gupta, D. (2022c). Mathematical Approach on Chipping Volume Estimation Generated During Rotary Ultrasonic Drilling for Float Glass. *Proceedings of the National Academy of Sciences, India Section A: Physical Sciences*, 92, 285–291.

Sharma, A., Jain, V., Gupta, D., & Babbar, A. (2020a). Review Study on Miniaturization. In *Advanced Manufacturing and Processing Technology* (1st ed., pp. 111–131). Boca Raton, FL: CRC Press.

Sharma, A., Kalsia, M., Uppal, A.S., Babbar, A., & Dhawan, V. (2022a). Machining of Hard and Brittle Materials: A Comprehensive Review. *Materials Today: Proceedings*, 50, 1048–1052.

Sharma, A., Kalsia, M., Uppal, A.S., Babbar, A., & Dhawan, V. (2022d). Machining of Hard and Brittle Materials: A Comprehensive Review. *Materials Today: Proceedings*, 50, 1048–1052.

Sharma, A., Kumar, V., Babbar, A., Dhawan, V., Kotecha, K., & Prakash, C. (2021b). Experimental Investigation and Optimization of Electric Discharge Machining Process Parameters Using Grey-Fuzzy-Based Hybrid Techniques. *Materials, Materials*, 14(19), 5820.

Sharma, A., Parikh, P.A., Roy, D., Joshi, K., & Trivedi, R. (2023c). Performance Evaluation of an Indigenously-Designed High-Performance Dynamic Feeding Robotic Structure Using Advanced Additive Manufacturing Technology, Machine Learning and Robot Kinematics. *International Journal on Interactive Design and Manufacturing (IJIDeM)*, 1–29.

Sharma, A., Sandhu, H.S., Goyal, D., Goyal, T., Jarial, S., & Sharda, A. (2023a). Sustainable Development in Cold Gas Dynamic Spray Coating Process for Biomedical Applications: Challenges and Future Perspective Review. *International Journal on Interactive Design and Manufacturing (IJIDeM)*, 1–17. https://doi.org/10.1007/s12008-023-01474-7

Sharma, V.K., Rana, M., Singh, T., Singh, A.K., & Chattopadhyay, K. (2021). Multi-Response Optimization of Process Parameters Using Desirability Function Analysis During Machining of EN31 Steel Under Different Machining Environments. *Materials Today: Proceedings*, 44, 3121–3126. https://doi.org/10.1016/j.matpr.2021.02.809

Shlykov, E.S., Ablyaz, T.R., & Oglezneva, S.A. (2020). Electrical Discharge Machining of Polymer Composites. *Russian Engineering Research*, 40, 878–879.

Singh, B.P., Singh, J., Bhayana, M., & Goyal, D. (2021). Experimental Investigation of Machining Nimonic-80A Alloy on Wire EDM Using Response Surface Methodology. *Metal Powder Report*, 76, 9–17. https://doi.org/10.1016/j.mprp.2020.12.001

Singh, B.P., Singh, J., Bhayana, M., Singh, K., & Singh, R. (2022). Experimental Examination of the Machining Characteristics of Nimonic 80-A Alloy on Wire EDM. *Materials Today: Proceedings*, 69, 291–296. https://doi.org/10.1016/j.matpr.2022.08.537

Singh, J., Singh, C., & Singh, K. (2023). Rotary Ultrasonic Machining of Advance Materials: A Review. *Materials Today: Proceedings*. https://doi.org/10.1016/j.matpr.2023.01.159

Sonbaty, E.I., Khashaba, U.A., & Machaly, T. (2004). Factors Affecting the Machinability of GFRP/Epoxy Composites. *Composite Structures*, 63, 329–338.

Sreejith, P.S., Krishnamurthy, R., Narayanasamy, K., & Malhothra, S.K. (1998). Machining Characteristics of Carbon/Phenolic Ablative Composites. *Journal of Materials Processing Technology*, 2, 503–507.

Srivatsan, T.S., & Bowden, D.M. (1992). Machining of Composite Materials. *Proceedings of the Machining of Composite Materials Symposium.* Chicago, IL: ASM/TMS Materials Week.

Suresh, J.S., Pramila Devi, M., & Sasidhar, M. (2017). Effect on Mechanical Properties of Epoxy Hybrid Composites Modified with Titanium Oxide (TiO_2) and Silicon Carbide (SiC). *International Journal of Engineering Science and Computing (IJESC)*, 7(10), 15193–15196.

Tagliaferri, V., Caprino, G., & Ditterlizzi, A. (1990). Effect of Drilling Parameters on the Finish and Mechanical Properties of GFRP Composites. *International Journal of Machine Tools and Manufacture*, 30, 77–84.

Tandon, S., Jain, V., Kumar, P., & Rajurkar, K. (1990). Investigations into Machining of Composites. *Precision Engineering*, 12(4), 227–238.

Taylor, R. (2000). Carbon Matrix Composites. In *Comprehensive Composite Materials* (pp. 387–426). Oxford: Pergamon.

Tonshoff, H.K., Lierse, T., & Inasaki, I. (1998). *Grinding of Advanced Ceramics, Machining of Ceramics and Composites* (Ed. Marcel Dekker, pp. 85–118). New York: Marcel Dekker.

Valino, A.D., Dizon, J.R.C., Espera, A.H., Chen, Q., Messman, J., & Advincula, R.C. (2019). Advances in 3D Printing of Thermoplastic Polymer Composites and Nanocomposites. *Progress in Polymer Science*, 98, 101162.

Velayudham, A., Krishnamurthy, R., & Soundarapandian, T. (2005). Evaluation of Drilling Characteristics of High Volume Fraction Fiber Glass Reinforced Polymeric Composite. *International Journal of Machine Tools and Manufacture*, 45, 399–496.

Wang, L., Lau, J., Thomas, E.L., & Boyce, M.C. (2011). Co-continuous Composite Materials for Stiffness, Strength, and Energy Dissipation. *Advanced Materials*, 23, 1524–1529.

Yahaya, R., Sapuan, S., Jawaid, M., Leman, Z., & Zainudin, E. (2014). Mechanical Performance of Woven Kenaf-Kevlar Hybrid Composites. *Journal of Reinforced Plastics and Composites*, 33, 2242–2254.

7 Optimization of Machining Parameter Using Electric Discharge Machining on Fabricated Aluminium-Based Metal Matrix Composite

Rajinder Kumar, Navdeep Singh, Harish Kumar Garg

7.1 INTRODUCTION

In the recent era of technology, the demands for metal matrix composite materials are rapidly increased due to their high corrosion-resistance property and good machineability. Due to this, it is mostly used in the field of aircraft, aerospace, and automotive industries. But it is very difficult to machine these composite materials by traditional machining methods (Gupta et al., 2014; Srivastava et al., 2014; Nag et al., 2018; Jha et al., 2014). The constituents of composites are called the individual material that makes up them. The advantages of metal matrix composite are having a lower coefficient of thermal expansion with high stiffness and specific strength and better fatigue resistance. Composites are not only used for their properties but they are also used to enhance their structural properties. The composite is defined as follows: "The composites are compound materials which differ from alloys by the fact that the individual components retain their characteristics but are so incorporated into the composite as to take advantage only of their attributes and not of their shortcomings" (Verma et al., 2013).

One of the most recently created materials, metal matrix composites are presently attracting the attention of industries due to their exceptional mechanical qualities. However, due to severe tool erosion, it is highly challenging to manufacture such materials using a conventional machine. Through the non-conventional machining technique, such an issue may be shorted out. Therefore, non-conventional machining is best suited to mill the aforementioned materials. Spark erosion machining (EDM) is the most used non-conventional machining method for such materials (Rizwee & Rao, 2021; Rizwee et al., 2021). A crucial non-traditional technology known as electric discharge

DOI: 10.1201/9781003327905-7

machining (EDM) is widely used to manufacture a variety of challenging materials, including heat-treated tool steels, composites, ceramics, carbides, and heat-resistant steels with convoluted shapes (Gupta & Kumar, 2021).

EDM is a well-liked non-traditional machining technique that can produce intricate forms. In EDM, the only requirement is that the workpiece is electrically conducting (Trzepiecinski et al., 2022; Maurya et al., 2022). Because copper has a superior surface polish, a higher MRR, a lower diametric overcut, and less electrode wear than other tool electrode materials, copper electrodes provide relatively low electrode wear for aluminium workpieces (Singh et al., 2004) . The significant localized stress caused by the high-temperature differential created at the inter- electrode gap led to material removal (Yadav et al., 2002). Therefore, modeling, simulation, and optimization are the three important issues for manufacturing a product. Out of these three approaches for any non-traditional machining, it is very essential to optimize the machining parameters. The optimization not only increases the utility for the economics of machining but also increases the product quality to a great extent. EDM is a precision machining process used to machine complex or simple geometries within parts and assemblies. The effects of control parameters such as pulse-on time, pulse-off time, and current on the material removal rate over the machining of metal matrix composites with regards to electric discharge machining were investigated, where crushed ash of mustard husk and SiC as reinforcement with the aid of Taguchi L18 orthogonal array. Two different types of tool electrodes such as copper and brass were used to conduct the comparative study (Choudhary, 2019).

Current, pulse-on, and pulse-off times, as well as the gap voltage, were used as input parameters, while MRR was set as the output parameter. The study found that voltage has a very small impact on the material's MRR. Additionally, it was shown that the MRR is mostly affected by the current and pulse-on time. MRR decreased as pulse-off time was lengthened (Singh & Maharana, 2020). The effect of current (I), voltage (V), and pulse on-time (T_{on}) on the cutting rate and surface finishing during machining Al 7075-B4C. Results found that beyond the mentioned control parameters possess a higher cutting rate and poor surface finish (Gopalakannan et al., 2012). It was discovered that when flushing pressure, supply current, and pulse duration rise, so does the rate of material removal. Additionally, the rate of tool wear increases proportionally as supply current increases and decreases in proportion to pulse-on time. Conducted studies using grey relational analysis on Al6061 reinforced with Al_2O_3. The Al6061-Al_2O_3 composite's MRR and surface roughness were mostly affected by the pulse current (Hourmand et al., 2019).

The Al 20Mg2Si composite was produced by melting a commercially available ADC12 alloy (Al 11.7Si-2Cu) using a 2 kg SiC crucible in an induction furnace. After melting, pure aluminium (99.7 wt %) and pure magnesium (99.9 wt %) were added to adjust the composite composition (Hourmand et al., 2019). The prediction accuracy of the ANN model is almost three times better than that of the response surface methodology. The percentage of error in the ANN prediction of MRR was found to be around 5.25%, while for the RSM model, it was more than 15% (Shandilya et al., 2013), optimizing the EDM of the 8% SiC/Al6061 composite. The work used Box-Behnken design (BBD) for the experimental design and response surface methodology (RSM) for the mathematical model. Current, pulse-on time, and duty cycle were the input parameters used. The material removal rate (MRR), electrode wear rate (EWR), and surface roughness of the machined specimen were all examined in this work. The current predominated the fluctuation in MRR (Srivastava et al., 2021). MRR increases as voltage rises from 40 V to 50 V. Additionally, when the voltage is raised from 50 to 60 volts, MRR first rises before beginning to fall (Singh et al., 2021). The environmental impact of hydrocarbon-based dielectric versus green dielectrics like water, dry EDM, and near-dry EDM has been examined, along with the sustainability of EDM (Kalyon & Fatatit, 2019; Evertz et al., 2006).

Using the multi-response function of the RSM approach, WEDM parameters like MRR and Kerf width were optimized. It was located that the main factors influencing material removal rate and kerf breadth are peak current, gap voltage, and duty cycle (Singh et al., 2021). It was also observed that a higher value of current causes frequent wire breakage during cutting (Bobbili et al., 2015). Chaudhari et al. (2020) worked on the surface analysis of shape memory alloy by WEDM. They found that defect-free and better surface finish obtained at optimized values of current, pulse-on time, and pulse-off time. For varying pulse on time and wire tension, the wire electrical discharge machining (EDM) of 6061 aluminium alloy was examined in terms of material removal rate, kerf/slit width, surface finish, and electrode wire wear. It appears that increased wire tension makes machining more stable, which results in less wire electrode wear and greater surface smoothness. Constant dielectric fluid supply pressure and specimen thickness were used (Pramanik et al., 2015). A fascinating recent development in stir casting is a two-step mixing process. In this process, the matrix material is heated above its liquidus temperature initially. It is then kept at a temperature halfway between the liquidus and solidus points to maintain its semi-solid state. Now add and combine the warmed reinforcement particles (Singla et al., 2009), developing a composite of aluminium 6063 reinforced with 5% SiC (30 mm in size). The outcome revealed that the fabricated MMC had increased toughness and strength. Further, the authors studied the effectiveness of MMC produced using EDM. The results concluded that all the input parameters have a positive effect on the MRR (Srivastava, 2020).

There is a fresh wave of change sweeping through the manufacturing sector. The manufacturing sector is no exception to the rule that technological progress and innovation go hand in hand. These additive and subtractive manufacturing developments are motivated by a variety of factors, including financial and ecological considerations. The future of manufacturing, including the functions of additive and subtractive methods, is laid bare by this study (Akhai & Rana, 2022; Babbar et al., 2020a, 2020b, 2020c, 2020d, 2021, 2022; Kalia et al., 2022; Khanduja et al., 2021; Kumar et al., 2021; Parikh et al., 2023; Prakash et al., 2021; Rampal et al., 2021; Rana & Akhai, 2022; Sharma et al., 2018a, 2018b, 2019a, 2019b, 2020a, 2020b, 2021a, 2021b, 2021c, 2022a, 2022b, 2022c, 2023a, 2023b; Sharma & Jain, 2020; Sharma V K et al., 2021; Singh et al., 2022, 2023).

The composite is produced by the stir casting method (SCM). To find out the SiC particles distribution in the Al matrix of the composites (as-cast), an optical microscope was used (Shehata et al., 2012). The process conditions, such as the stirring speed and the fixed temperature, can have an impact on the mechanical properties when this approach is used (Ibrahim et al., 2022). The machining properties of an aluminium-based metal matrix composite were examined by utilizing the EDM method and aluminium powder (average particle size of 15 μm) in a kerosene-based dielectric medium and discovered that the MRR is improved when aluminium powder is added to the dielectric medium (Talla et al., 2015). Adding powder to the dielectric fluid to better the machining process could affect or improve the effectiveness of the fluid. The effects depend on the type of powder and its concentration (Abukhshim et al., 2006). De-ionized water is used as a dielectric fluid, and brass wire of 0.25 mm diameter is used as an electrode. At first, we conducted experiments to identify the parameters that affect material removal rate and surface roughness. The analysis was carried out using the statistical analysis of variance (ANOVA) technique (Rani et al., 2017). The influence of four machining parameters (duty cycle, pulse duration, discharge current, and flushing pressure) and two material parameters (mesh size and weight fraction of silicon carbide in the composite) on the material removal rate has been analyzed by repeated experiments carried out on the workpieces. The average values of the key machining parameters were obtained in the present study (Puhan et al., 2013). Electrode wear rate is high in micro EDM due to short pulse duration. They investigated how materials used for the tool electrodes affect wear and the surface formation process (Uhlmann & Roehner, 2008). In comparison to typical manufacturing methods, this results in less tool wear. The material is removed in the presence of the dielectric medium by creating carefully controlled electrical discharges between the tool and the workpiece (Wasif et al., 2022).

7.2 EXPERIMENTAL DETAIL

7.2.1 Fabrication of Work Materials

Work materials in this investigation are the aluminium (Al6061) based MMC which was fabricated with the aid of the stir casting technique. The chemical composition of Al6061 is given in Table 7.1. Silicon carbide particles (SiC) and alumina (Al2O3) act as a reinforcement and are the major constitute with the proportion of Al:SiC:Al2O3 = 80:16:4 respectively.

Firstly, the aluminium was converted to molten form by placing the solid materials into the furnace. On the flip side, the reinforcement materials were preheated at 630°C for three hours separately in another furnace (Kumar et al., 2016). The MMCs were prepared under the temperature of 750°C, and molten metal with reinforcement was mixed with the help of a motor stirrer. The speed of the motor is maintained at 500 RPM for three hours to become a homogeneous mixture. Further, the mould

TABLE 7.1
Chemical Composition of Al6061

Element	Al	Mg	Cu	Zn	Si	Fe	Mn	Cr	Others	Ti
Max Weight (%)	Rest	0.81–1.21	0.15–0.40	Max 0.25	0.41–0.82	Max 0.71	Max 0.15	0.04–0.35	0.05	Max 0.15

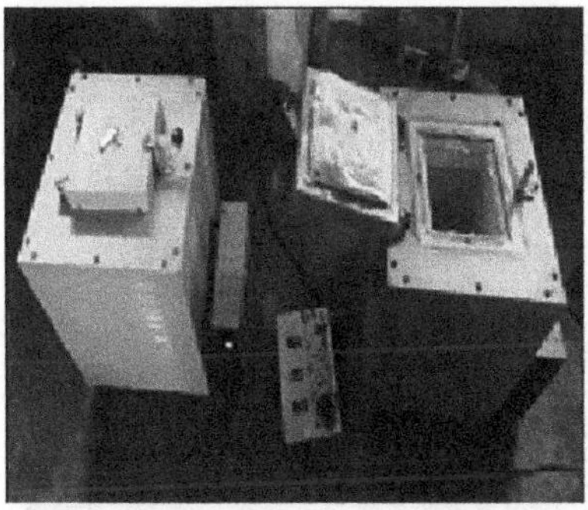

FIGURE 7.1 Pictorial view of equipment used during the casting process.

was preheated at 500°C temperature for 30 minutes and then obtained molten metal is poured into the mould of the desired shape to obtain uniform solidification and avoid the oxidation during the pouring of molten metal in it. However, the solidification of molten metal carried under room temperature. After the turning operation, workpieces were divided into a number of pieces having a diameter of 44 mm and a height of 15 mm.

7.2.2 Property Analysis of Al/(SiCp + Al2O3p)—MMC

Hybrid metal matrix composites are a diverse class of materials in which all constituents maintain their identity, as they do not dissolve or melt in each other and act in such a way that a new material results whose properties are better than the sum of their constituents. The addition of graphite in Al/SiC MMC put a significant effect on the properties of the casted composite specimens. Following are the tests performed to check the mechanical properties of the cast specimen.

7.3 TENSILE TEST

A tensile test on the prepared sample of length 200 mm and diameter 10 mm was conducted using comprised UTM machine of capacity 400 KN, make mechatronics. A tensile test was performed three times on material of the same composition and dimensions to have accurate results (Table 7.2). Specimen testing before and after tensile testing is shown in Figure 7.2 and Figure 7.3.

TABLE 7.2
Tensile Strength of the Aluminium/(SiCp + Al2O3) Hybrid Composite

Material	Sample 1 (MPa)	Sample 2 (MPa)	Sample 3 (MPa)	Avg. Tensile Strength (MPa)	Percentage Elongation
Al/16% SiCp + 4% Al2O3p	260	262	256	259.33	3.8

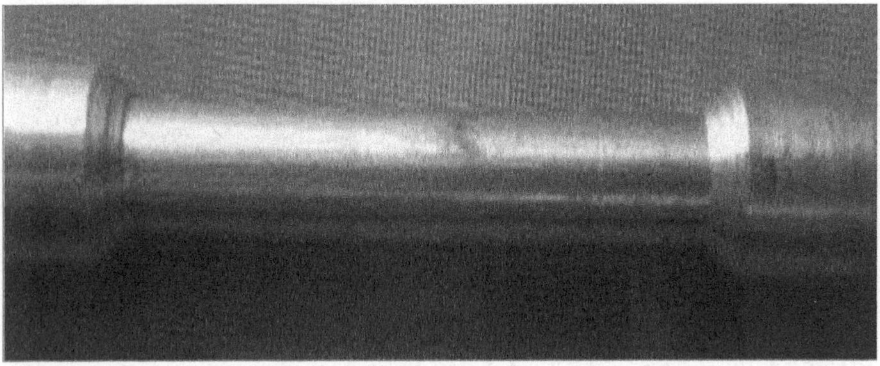

FIGURE 7.2 Specimen of MMC in desire shape before tensile test.

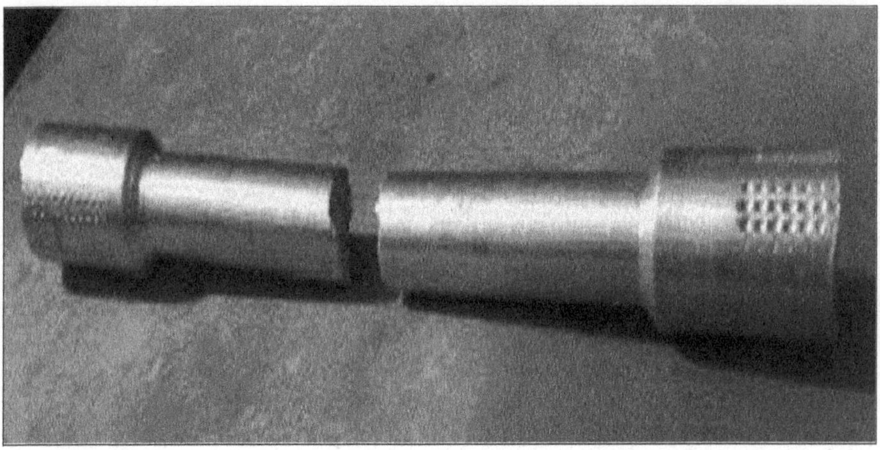

FIGURE 7.3 Specimen of MMC in desire shape after tensile test.

7.4 DENSITY TEST

Density is a measure of the "compactness" of matter within a substance. The hydrometer measures density directly. An object that is less dense than a liquid will float in that liquid density to a depth such that the mass of the object submerged equals the mass of the liquid displaced (Archimedes' principle).

Density measurements were carried out on the base metal and reinforced samples using the Archimedes' principle. This method of density measurement simply involves weighing the sample in air and another fluid of known density. Application of Archimedes' principle leads to the following expression for the density (ρmmc) of the composites:

$$\rho mmc = m1/(m1—m2)\ \rho w\ g/cm^3$$

Where m_1 is the mass of the composite sample in air, m_2 is the mass of the same composite sample in distilled water, and ρw is the density of the distilled water. The density of distilled water at 20°C is 998 kg/m^3.

7.4.1 Procedure Followed to Measure Density

Weigh and record the mass of the composite sample of cylindrical shape using a precise weighing machine with 0.001g least count. Calculate the volume of the composite cylinder by measuring (in cm) the height (h) and diameter (d) of the metal cylinder and then applying the formula: volume (cc) = h x 0.785d2. Also, measure the volume of the composite cylinder by displacement of water in a 50 ml graduated cylinder. Calculate the density of the composite cylinder for each method of measuring volume.

Area of circle ¼ πd2, where d = diameter and π = 3.14159 volume of a cylinder = area of base x height. Using this method, the densities of the base metal and hybrid composites were measured. The density of stir cast Al/16wt%SiC/4wt%Al2O3-MMC. Table 7.3 represent the value of density which is calculated by using the density measurement apparatus. From the sample, density is measured at three different times and the average value is calculated by the sum of the measured value divided by the number of tests performed.

7.5 HARDNESS TEST

During microhardness testing, a Vickers (DPH) diamond indenter is pressed into the material's surface with a penetrator and a light load of up to 1,000 grams. The result of applying the load with a penetrator is an indent or permanent deformation of the material surface caused by the shape of the indenter. Vickers hardness test methods use specific measurements from the indent, in conjunction with formulas to calculate material hardness. Accurate measurement of the resulting indentation requires the use of a special microhardness-testing microscope because the indents are so small. For this study, test specimen of Al/16wt%SiC/4wt%Al2O3-MMC composition is used to check microhardness by a highly precised microhardness tester by Mitutoyo (Figure 7.4) at NITTTR, Chandigarh. Before the test specimen was polished from both sides using diamond paste, a load of 200 g was applied for ten seconds. A square indentation was formed on the workpiece, and by measuring the distance between opposite corners of the indentation, the final reading for hardness was on the screen. Tests were performed three times at different locations to have the

TABLE 7.3
Density of Aluminium/(SiCp + Al2O3) Hybrid Composite

Material	Test 1 (g/cm3)	Test 2 (g/cm3)	Test 3 (g/cm3)	Avg. Density
Al/16% SiCp + 4 % Al2O3 p	2.82	2.85	2.80	**2.82 g/cm3**

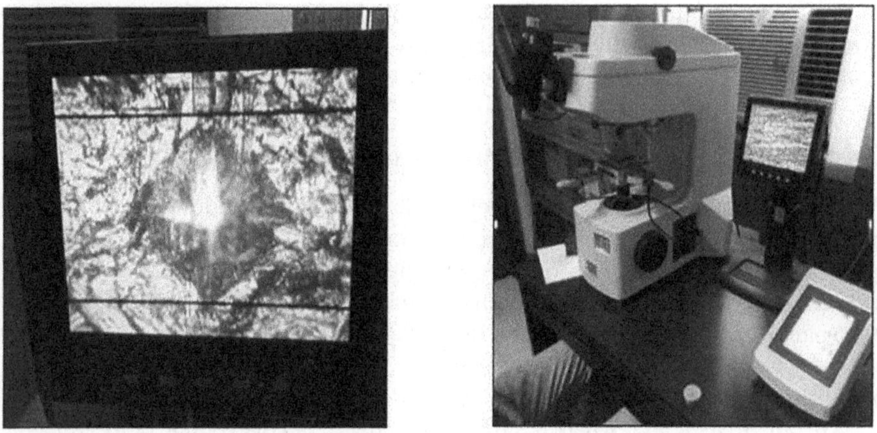

FIGURE 7.4 Microhardness tester (Mitutoyo).

TABLE 7.4
Microhardness of Aluminium/(SiCp + Al2O3) Hybrid Composite

Material	Position 1 (HV)	Position 2 (HV)	Position 3 (HV)	Avg. Hardness (HV)
Al/16wt%SiC/4wt%Al2O3-MMC	118	119	123	120

correct value of hardness. The average microhardness of the tested sample is 120 HV (Table 7.4), whereas the hardness of the base metal is 107 HV (ASM material data).

7.6 EXPERIMENTAL CONDITIONS AND PROCEDURE

In this analysis, the experiment was taken into account during the process of electric discharge machining (CMAX S645) of metal matrix composite. It is an advanced CNC type of sinker EDM machine having features with three-dimensional (3D) mode motion machining. The six axes are controllable with C, A, and B axes that serve as optional electrode heads.

A 0.025 mm gap is maintained between the workpiece and tool. Moreover, for automation, it can be connected to the robot's hand, and the electrodes can be replaced by the hand of the robot. For the smooth function of the process consideration of dielectric fluid, EDM oil (flash point = 94°C, specific gravity = 0.763) was taken.

A copper cylinder bar that has a diameter of 12 mm was used in the experiment. Moreover, the use of a digital weighing machine (least count = 0.1 mg) after each run to measure the surface roughness and material removal rate was measured with a surface roughness device having a model INSIZE ISR-S400. Figure 7.5 shows the detail of the electric discharge machine and thc operation performed by it.

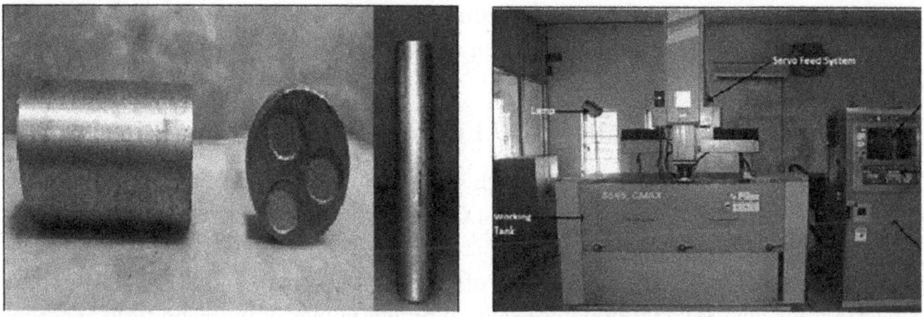

FIGURE 7.5 Casted, machined workpiece, copper tool, and electric discharge machine.

TABLE 7.5
Experimental Parameters and Their Levels

S. No.	Parameter	Notation	Unit	Level		
				−1	0	1
1	Pulse-on time (T_{on})	A	μs	30	60	90
2	Pulse-off time (T_{off})	B	μs	30	60	90
3	Current (I)	C	A	6	7	8
4.	Voltage (V)	D	V	10	12	14

To attain the best staging of the EDM process and to ensure that the results are predictable, parametric design is necessary and involves characterizing multiple process responses, such as surface finish, material removal rate, surface integrity and tool wear rate, heat affected zone, etc., concerning different machining parameters, like peak current, pulse-on time, pulse-off time, gap voltage, duty factor, dielectric flushing pressure, dielectric fluid, etc. The study found that different combinations of EDM process parameters are needed to achieve a higher MRR and lower SR. In this investigation, pulse-on time, pulse-off time, peak current, and gap voltage are selected as numeric factors. Table 7.5 shows the experiment parameter and their level.

7.7 RESULTS AND DISCUSSIONS

This study presents the effects of different processing parameters on the surface roughness and removal rate of an electrical discharge machining (EDM) process. Pulse-on time, pulse-off time, current, and voltage were varied, while EDM experiments were conducted to analyze how they affect the response of surface roughness and material removal rate. In order to identify the main parameters (significant) as well as the interaction effect of process parameters, analysis of acquired data is important. Design expert 7.0.0® software was used to check the adequacy of the model. To check the acceptability of the model, ANOVA technique was used, as lack of fit and model summary statistics. Single objective optimization is done using design expert 7.0.0® software. The following results have been drawn from the experiments.

7.7.1 Parametric Interaction Effect on Surface Roughness Height

EDM is known for its accuracy and high surface quality. A good response output is an interaction effect of input parameters. Given next are some graphs for the significant interaction effect.

Figure 7.6 represent the interaction effect of pulse-on time and pulse-off time, reflecting that surface quality deteriorates at higher values of pulse-on time, whereas with the increase in pulse-off time, there is a significant improvement in surface quality. From the plot, it may experimentally prove that with the interaction of both pulse-on time and pulse-off time, a lower value of pulse-on time and a higher value of pulse-off time may lead to higher surface quality.

Surface roughness decreases, whereas pulse-off time goes higher at a constant value of pulse-on time. At higher values of pulse-off time and lower values of pulse-on time, the surface roughness was found to be decreased. Whereas a lower value of both pulse-on time and current leads to better surface quality. From Figure 7.7, surface roughness has been found to be high at a higher value of both pulse-on time and current. Whereas, surface quality is best at lower values of both current (I) and pulse-on time (T_{on}). Lower the duration of spark time (pulse-on time) and low intensity of the energy (current) result in lower surface roughness.

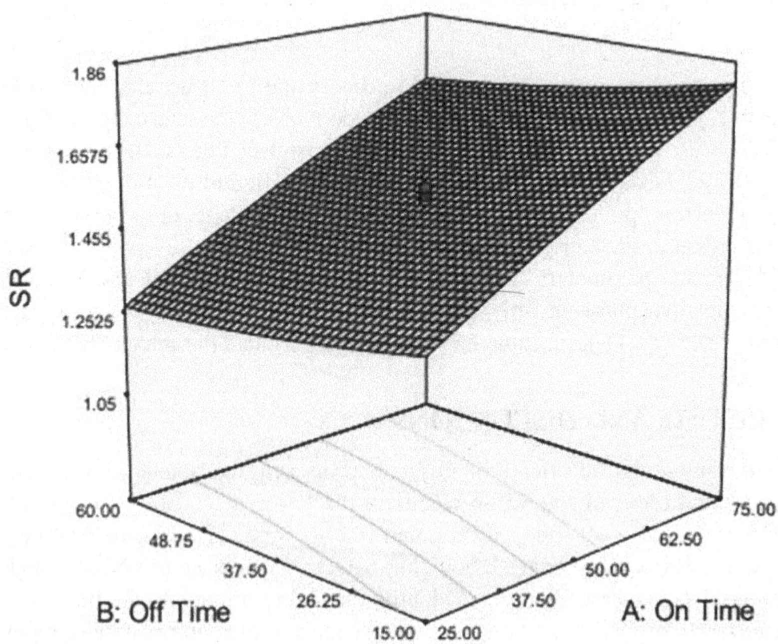

FIGURE 7.6 Interaction effect of pulse-off time and pulse-on time on surface roughness height.

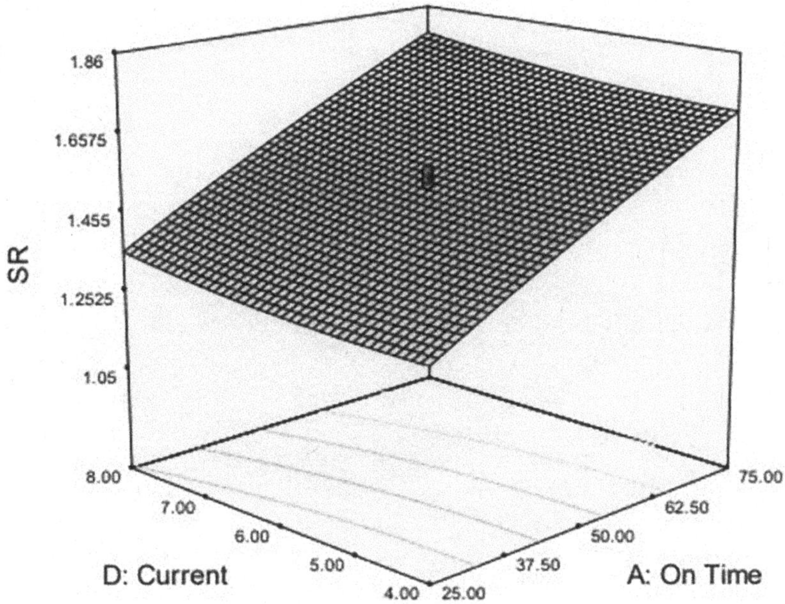

FIGURE 7.7 Interaction effect of current and pulse-on time on surface roughness height.

Surface roughness is also a strong function of voltage (V) and pulse-off time (T_{off}). From Figure 7.8, it is observed that surface roughness is higher at the increased value of pulse-off time and raised value of voltage. With the interaction of both pulse-off time and voltage, a lower value of pulse-off time and voltage may lead to lower surface roughness. Surface roughness value is minimum at a low value of voltage and current, Figure 7.9 (i.e. at 3 V and 5.5 A). Thereafter, the roughness height increases on an increase in current and voltage. High voltage and higher value of current (intensity of energy discharge) result in high surface roughness.

7.8 PARAMETRIC INTERACTION EFFECT ON MRR

There is a notable effect of the interaction of process parameters on MRR.

From Figure 7.10, it is clear that MRR (material removal rate) is increasing with an increase in pulse-on time and decreases with an increase in pulse-off time. Although with the interaction of both pulse-off time and pulse-on time, the pulse-off time has little impact and pulse-on time has a greater impact. Also, pulse-off time has a little slope as compared to pulse-on time with respect to MRR. The high impact of the pulse-on time as compared to the pulse-on time over the MRR is only caused by the high time interval of spark produced during the machining.

Figure 7.11 shows the interaction between the pulse-on time (T_{on}) and voltage (V) over the MR (material removal) rate. From the graph, it is clearly observed that as the pulse-on time as well as voltage increases, the material removal rate increases,

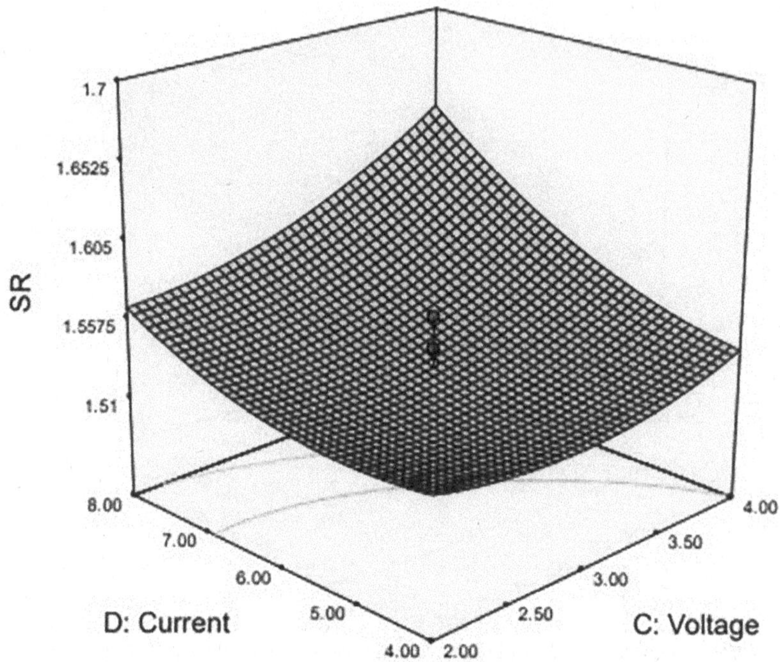

FIGURE 7.8 Interaction effect of pulse-off time and voltage on surface roughness.

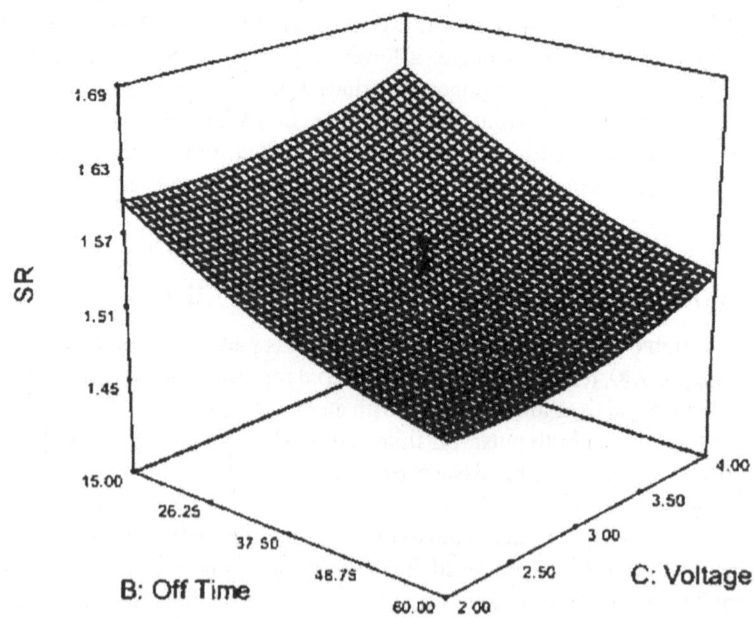

FIGURE 7.9 Interaction effect of current and voltage on surface.

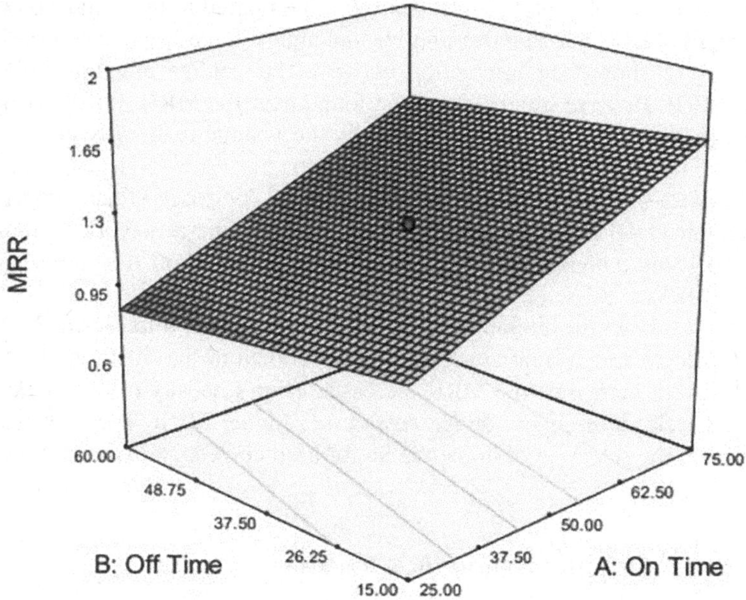

FIGURE 7.10 Interaction effect of pulse-on time and pulse-off time on MRR.

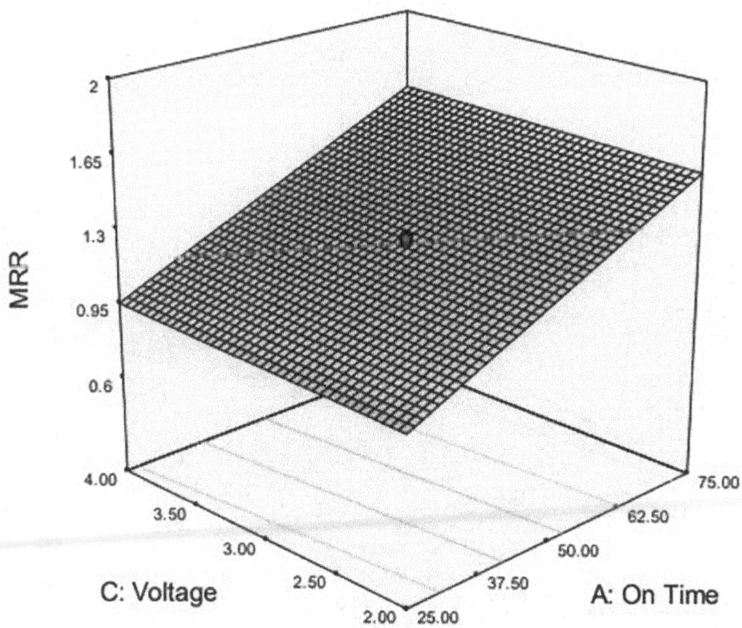

FIGURE 7.11 Interaction effect of pulse-on time and voltage on MRR.

where the pulse-on time has a higher impact as compared to the voltage on MRR. For the higher MRR, a high pulse-on time and high voltage range may be selected.

Figure 7.12 shows the interaction between the voltage and pulse-off time over the MRR. Reverse trends have been found over the MRR. Pulse-off time is inversely proportional to the MRR, whereas the voltage is directly proportional to MRR.

As the pulse-off time increases, the spark interval decreases, which further leads to a lower rate of deformation of the materials. With the interaction of both pulse-off time and current, a higher value of voltage and lower pulse-off time are suggested for higher MRR.

Figure 7.13 shows the interaction effect of the current and voltage over the MRR. Both the voltage and current are directly proportional to the MRR. As both voltage and current increases, the MRR increases. High intensity of the spark (peak current) as well as the high voltage results in a higher MRR. For high material removal rate, the selection of higher value for both current and voltage have been suggested.

7.8.1 Single Objective Optimization for SR

Surface quality is the most prominent factor in product manufacturing. As far as single objective optimization is concerned, this can be achieved at low MRR given next is a predicted set of parametric values for the optimal value of surface quality.

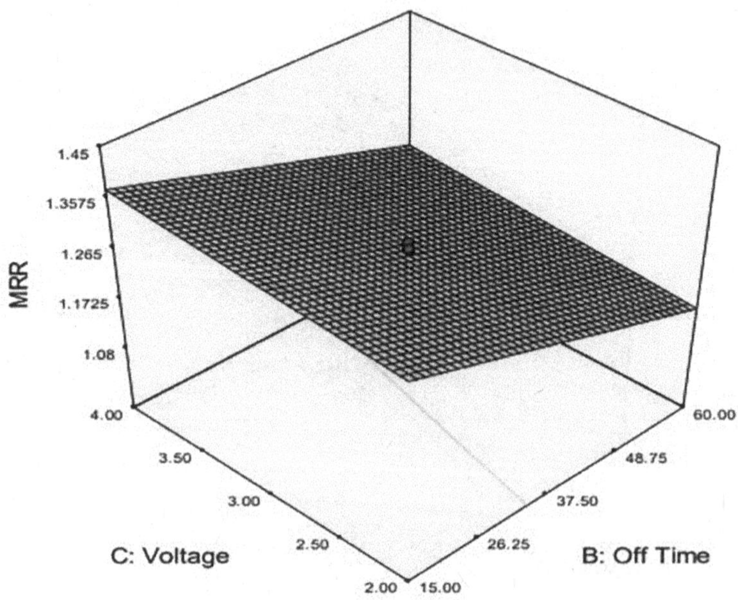

FIGURE 7.12 Interaction effect of current and voltage on MRR.

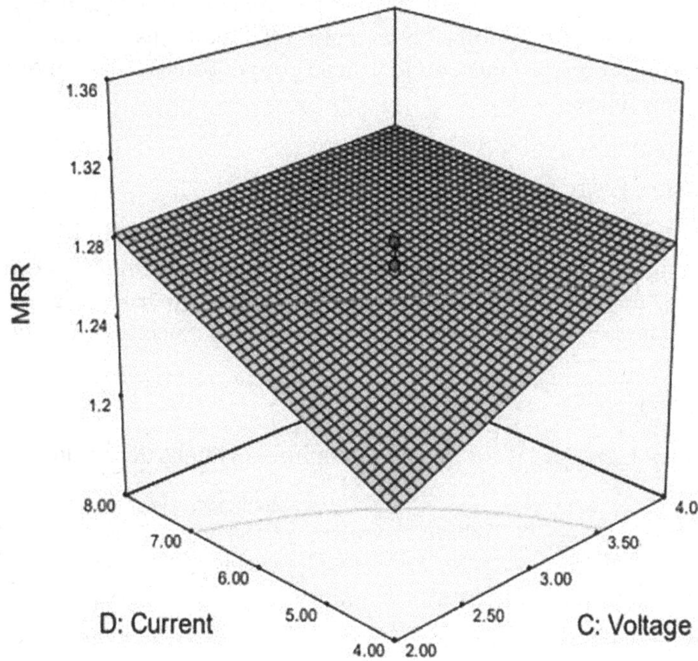

FIGURE 7.13 Interaction effect of current and pulse-off time on MRR.

TABLE 7.6
Single Objective Optimization for SR Using Desirability Approach

Constraints						
Lower Name	**Upper Goal**	**Lower Limit**	**Upper Limit**	**Weight**	**Weight**	**Importance**
On time	is in range	25	75	1	1	3
Off time	is in range	15	60	1	1	3
Voltage	is in range	2	4	1	1	3
Current	is in range	4	8	1	1	3

Sr. No	On Time (μs)	Off Time (μs)	Voltage (V)	Current (A)	SR (μm)	Desirability
1	25.00	60.00	2.12	6.06	1.26027	0.764 **(Selected)**
2	25.00	60.00	2.03	6.12	1.26034	0.764
3	25.02	60.00	2.13	6.00	1.2605	0.763
4	25.00	60.00	2.04	6.37	1.2606	0.763

Surface roughness height of 1.26 μm can be achieved as predicted by the desirability approach at pulse-on time 25 μs, pulse-off time 60 μs, voltage ≈ 2 V, and current ≈ 6 A as process parameters. To validate the predicted results validity experimentation is as follows.

7.8.2 Single Objective Optimization for MRR

EDM is a slow process. The industry is always interested to have maximum production (i.e. maximum production per day as far as rough machining is concerned). Therefore, the optimal set of EDM parameters is important to have a high MRR. The desirability approach in RSM is used to optimize EDM parameters for desired MRR.

TABLE 7.7
Confirmatory Experiment for Surface Roughness Height (Ra in μm)

Sr No	On Time (μs)	Off Time (μs)	Voltage (V)	Current (A)	Predicted SR (μm)	Experimental SR (μm)	Percentage Error
1	25	60	2	6	1.26	1.18	-6.34

TABLE 7.8
Single-Objective Optimization for MRR

		Constraints				
Lower	Upper	Lower	Upper			
Name	Goal	Limit	Limit	Weight	Weight	Importance
On time	is in range	25	75	1	1	3
Off time	is in range	15	60	1	1	3
Voltage	is in range	2	4	1	1	3
Current	is in range	4	8	1	1	3

S. No.	On time	Off time	Voltage	Current	MRR (g/min)	Desirability
1	75.00	15.05	3.89	8.00	1.71042	0.828
2	75.00	15.01	3.39	8.00	1.71039	0.828
3	75.00	15.00	2.79	8.00	1.71027	0.828
4	75.00	15.00	2.33	7.98	1.70965	0.827
5	74.98	15.00	2.10	8.00	1.70962	0.827

TABLE 7.9
Confirmatory Experiment for MRR

S. No.	On Time	Off Time	Voltage	Current	Predicted MRR	Experimental MRR	Percentage Error
1	75	15	4	8	1.711	1.801	5.26

The predicted MRR of ≅ 1.711 g/min can be achieved by on time 75 μs, off time 15 μs, voltage 3.89 V, and current 8 A by the copper tool.

7.9 CONCLUSION AND FUTURE OUTLOOK

- Tensile strength increases with the addition of alumina and silica as reinforcement. With the addition of reinforcement into the base material, the average tensile strength of the composite was observed to be 259.33 MPa with 3.8% of elongation.
- Tensile strength has been recorded highest for five min of stirring time. Further increase in stirring time will lead to a partial solidification of aluminium, and hence, tensile strength starts decreasing.
- Optimal value of MRR (1.80 g/min) has been achieved at, 8 A, 4.0 V, 75 μs, and 15 μs as peak current, gap voltage, pulse-on time, and pulse-off time, respectively.
- A correlation between SR and the frequency of pulse on time has been observed. Low frequencies result in a negligible SR, while high frequencies result in a high SR. The optimal value of SR (1.14 μm) has been achieved at 25 μs, 60 μs, 2 V, and 6 A as a pulse-on time, pulse-off time, gap voltage, and peak current respectively.

7.10 FUTURE SCOPE

- The size of the graphite particles also affects the material's wear resistance. The particle size employed in the current investigation is 200 mesh (avg. 75 m) for both SiC and alumina (Al2O3). To evaluate the machinability studies of hybrid MMC, the size and weight percentage of the reinforcement particles can be changed.
- In the current work, RSM has been applied to the desirability strategy to optimize the response variables. Other optimization methods exist, including artificial neural networks (ANN) and artificial bee colonies (ABC). The possibility of optimizing the machining settings could be investigated using genetic algorithms (GA), particle swarm optimization (PSO), instructor learning-based optimization, etc.
- Using a different set of electrodes to adjust the process parameters during EDM could lower the final product's cost.

REFERENCES

Abukhshim, N.A., Mativenga, P.T., & Sheikh, M.A. (2006). Heat generation and temperature prediction in metal cutting: A review and implications for high-speed machining. *International Journal of Machine Tools and Manufacture*, 46(7–8), 782–800.

Akhai, S., & Rana, M. (2022). Taguchi-based grey relational analysis of abrasive water jet machining of Al-6061. *Materials Today: Proceedings*, 65, 3165–3169. https://doi.org/10.1016/j.matpr.2022.05.361

Babbar, A., Jain, V., Gupta, D., Prakash, C., & Sharma, A. (2020a). Fabrication and machining methods of composites for aerospace applications. In *Characterization, Testing, Measurement, and Metrology* (1st ed., pp. 109–124). Boca Raton, FL: CRC Press.

Babbar, A., Jain, V., Gupta, D., Prakash, C., Singh, S., & Sharma, A. (2020b). Effect of process parameters on cutting forces and osteonecrosis for orthopedic bone drilling applications. In *Characterization, Testing, Measurement, and Metrology* (1st ed., pp. 93–108). Boca Raton, FL: CRC Press.

Babbar, A., Jain, V., Gupta, D., Prakash, C., Singh, S., & Sharma, A. (2020d). 3D bioprinting in pharmaceuticals, medicine, and tissue engineering applications. In *Advanced Manufacturing and Processing Technology* (1st ed., pp. 147–161). Boca Raton, FL: CRC Press, 2021.

Babbar, A., Jain, V., Gupta, D., & Sharma, A. (2020c). Fabrication of microchannels using conventional and hybrid machining processes. In *Non-Conventional Hybrid Machining Processes* (1st ed., pp. 37–51). Boca Raton, FL: CRC Press.

Babbar, A., Rai, A., & Sharma, A. (2021). Latest trend in building construction: Three-dimensional printing. *Journal of Physics: Conference Series*, 1950, 012007.

Babbar, A., Sharma, A., & Singh, P. (2022). Multi-objective optimization of magnetic abrasive finishing using grey relational analysis. *Materials Today: Proceedings*, 50, 570–575.

Bobbili, R., Madhu, V., & Gogia, A.K. (2015). Modelling and analysis of material removal rate and surface roughness in wire-cut EDM of armour materials. *Engineering Science and Technology, an International Journal*, 18(4), 664–668.

Chaudhari, R., Vora, J.J., Patel, V., López de Lacalle, L., & Parikh, D. (2020). Surface analysis of wire-electrical- discharge-machining-processed shape-memory alloys. *Materials*, 13(3), 530.

Choudhary, R. (2019). Optimization of machining parameters in EDM of AL 6061 15% SIC 6% MHA composite by Taguchi method. *International Journal of Engineering, Science and Mathematics*, 8(2), 49–57.

Evertz, S., Dott, W., & Eisentraeger, A. (2006). Electrical discharge machining: Occupational hygienic characterization using emission-based monitoring. *International Journal of Hygiene and Environmental Health*, 209(5), 423–434.

Gopalakannan, S., Senthilvelan, T., & Ranganathan, S. (2012). Modeling and optimization of EDM process parameters on machining of Al 7075-B4C MMC using RSM. *Procedia Engineering*, 38, 685–690.

Gupta, A., & Kumar, H. (2021). Optimization of EDM process parameters: A review of technique, process, and outcome. *Advances in Manufacturing and Industrial Engineering*, 981–996.

Gupta, P., Kumar, D., Parkash, O., & Jha, A.K. (2014). Effect of sintering on wear characteristics of Fe-Al2O3 metal matrix composites. *Proceedings of the Institution of Mechanical Engineers, Part J: Journal of Engineering Tribology*, 228(3), 362–368.

Hourmand, M., Sarhan, A.A., Farahany, S., & Sayuti, M. (2019). Microstructure characterization and maximization of the material removal rate in nano-powder mixed EDM of Al-Mg2Si metal matrix composite—ANFIS and RSM approaches. *The International Journal of Advanced Manufacturing Technology*, 101(9), 2723–2737.

Ibrahim, A.F., Singal, A.H., & Al Kareem Noori, D.A. (2022). Investigation of material removal rate and surface roughness during electrical discharge machining on Al (6061)-5% SiC-10% B4C hybrid composite. *Metallurgical and Materials Engineering*, 28(1), 47–60.

Jha, P., Gupta, P., Kumar, D., & Parkash, O. (2014). Synthesis and characterization of Fe—ZrO2 metal matrix composites. *Journal of Composite Materials*, 48(17), 2107–2715.

Kalia, G., Sharma, A., & Babbar, A. (2022). Use of three-dimensional printing techniques for developing biodegradable applications: A review investigation. *Materials Today: Proceedings*, 62, 346–352.

Kalyon, A., & Fatatit, A.Y. (2019). The environmental impact of electric discharge machining. *International Journal of Engineering Science and Application*, 3(3), 123–129.

Khanduja, P., Bhargave, H., Babbar, A., Pundir, P., & Sharma, A. (2021). Development of two-dimensional plotter using programmable logic controller and human machine interface. *Journal of Physics: Conference Series*, 1950, 012012.

Kumar, H., Kumar, R., & Manna, A. (2016). Effects of electrode configuration on MRR and EWR during electric discharge machining of Al/10wt% SiCp-MMC. *International Journal of Machining and Machinability of Materials*, 18(1–2), 54–76.

Kumar, S., Sudhakar, R.P., Goyal, D., & Sehgal, S. (2021). Process modelling for machining Inconel 825 using cryogenically treated carbide insert. *Metal Powder Report*, 76, 66–74. https://doi.org/10.1016/j.mprp.2020.06.001

Maurya, S.K., Susheel, C.K., & Manna, A. (2022). Experimental investigation of wire EDM parameters during machining of fabricated hybrid Al/(SiC+ ZrO2+ NiTi)-MMC. *Advances in Materials and Processing Technologies*, 141.

Nag, A., Srivastava, A.K., Dixit, A.R., Mandal, A., Das, A.K., & Tiwari, T. (2018). Surface integrity analysis of wire-EDM on in-situ hybrid composite A359/Al2O3/B4C. *Materials Today: Proceedings*, 5(11), 24632–24641.

Parikh, P., Sharma, A., Trivedi, R., Roy, D., & Joshi, K. (2023). Performance evaluation of an indigenously-designed high-performance dynamic feeding robotic structure using advanced additive manufacturing technology, machine learning and robot kinematics. *International Journal on Interactive Design and Manufacturing (IJIDeM)*, 1–29.

Prakash, C., Kumar, V., Mistri, A., Sharma, A., Uppal, A.S., Babbar, A., & Pathri, B.P. (2021). Investigation of functionally graded adherents on failure of socket joint of FRP composite tubes. *Materials*, 14, 6365.

Pramanik, A., Basak, A.K., Islam, M.N., & Littlefair, G. (2015). Electrical discharge machining of 6061 aluminium alloy. *Transactions of Nonferrous Metals Society of China*, 25(9), 2866–2874.

Puhan, D., Mahapatra, S.S., Sahu, J., & Das, L. (2013). A hybrid approach for multi-response optimization of non- conventional machining on AlSiCp MMC. *Measurement*, 46(9), 3581–3592.

Rampal, R., Goyal, T., Goyal, D., Mittal, M., Dang, R.K., & Bahl, S. (2021). Magneto-rheological abrasive finishing (MAF) of soft material using abrasives. *Materials Today: Proceedings*, 45, 51140–5121. https://doi.org/10.1016/j.matpr.2021.01.629

Rana, M., & Akhai, S. (2022). Multi-objective optimization of Abrasive water jet Machining parameters for Inconel 625 alloy using TGRA. *Materials Today: Proceedings*, 65, 3205–3210. https://doi.org/10.1016/j.matpr.2022.05.374

Rani, M.G., Rao, C.V., & Kotaiah, K.R. (2017). Experimental investigation on optimization of the controlling factors for machining al 6061/mos2 metal matrix composites with wire EDM. *International Journal of Applied Engineering Research*, 12(22), 12023–12028.

Rizwee, M., & Rao, P.S. (2021). Analysis & optimization of parameters during EDM of aluminium metal matrix composite. *Journal of University of Shanghai for Science and Technology*, 23(3), 218–223.

Rizwee, M., Rao, P.S., & Khan, M.Y. (2021). Recent advancement in electric discharge machining of metal matrix composite materials. *Materials Today: Proceedings*, 37, 2829–2836.

Shandilya, P., Jain, P.K., & Jain, N.K. (2013). RSM and ANN modeling approaches for predicting average cutting speed during WEDM of SiCp/6061 Al MMC. *Procedia Engineering*, 64, 767–774.

Sharma, A., Babbar, A., Jain, V., & Gupta, D. (2018b). Enhancement of surface roughness for brittle material during rotary ultrasonic machining. *MATEC Web of Conferences*, 249, 01006.

Sharma, A., Babbar, A., Tian, Y., Pathri, B.P., Gupta, M., & Singh, R. (2022a). Machining of ceramic materials: A state-of-the-art review. *International Journal on Interactive Design and Manufacturing (IJIDeM)*, 1–21. https://doi.org/10.1007/s12008-022-01016-7

Sharma, A., Fidan, I., Huseynov, O., Ali, M.A., Alkunte, S., Rajeshirke, M., Gupta, A., Hasanov, S., Tantawi, K., Yasa, E., Yilmaz, O., Loy, J., & Popov, V. (2023b). Recent inventions in additive manufacturing: Holistic review. *Inventions*, 8(4), 103. https://doi.org/10.3390/inventions8040103

Sharma, A., Grover, V., Babbar, A., & Rani, R. (2020b). A trending nonconventional hybrid finishing/machining process. In *Non-Conventional Hybrid Machining Processes* (1st ed., pp. 79–93). Boca Raton, FL: CRC Press.

Sharma, A., & Jain, V. (2020). Experimental investigation of cutting temperature during drilling of float glass specimen. In *IOP Conference Series: Materials Science and Engineering* (Vol. 715, No. 1, p. 012050). IOP Publishing. https://doi.org/10.1088/1757-899X/715/1/012050

Sharma, A., Jain, V., & Gupta, D. (2018a). Characterization of chipping and tool wear during drilling of float glass using rotary ultrasonic machining. *Measurement*, 128, 254–263.

Sharma, A., Jain, V., & Gupta, D. (2019a). Comparative analysis of chipping mechanics of float glass during rotary ultrasonic drilling and conventional drilling: For multi-shaped tools. *Machining Science and Technology*, 23, 547–568.

Sharma, A., Jain, V., & Gupta, D. (2019b). Multi-shaped tool wear study during rotary ultrasonic drilling and conventional drilling for amorphous solid. *Proceedings of the Institution of Mechanical Engineers, Part E: Journal of Process Mechanical Engineering*, 233, 551–560.

Sharma, A., Jain, V., & Gupta, D. (2021a). Effect of pre and post tempering on hole quality of float glass specimen: For rotary ultrasonic and conventional drilling. *Silicon*, 13, 2029–2039.

Sharma, A., Jain, V., & Gupta, D. (2022b). Mathematical approach on chipping volume estimation generated during rotary ultrasonic drilling for float glass. *Proceedings of the National Academy of Sciences, India Section A: Physical Sciences*, 92, 285–291.

Sharma, A., Jain, V., Gupta, D., & Babbar, A. (2020a). A review study on miniaturization. In *Advanced Manufacturing and Processing Technology* (1st ed., pp. 111–131). Boca Raton, FL: CRC Press.

Sharma, A., Kalsia, M., Uppal, A.S., Babbar, A., & Dhawan, V. (2022c). Machining of hard and brittle materials: A comprehensive review. *Materials Today: Proceedings*, 50, 1048–1052.

Sharma, A., Kumar, V., Babbar, A., Dhawan, V., & Kotecha, K. (2021b). Experimental investigation and optimization of electric discharge machining process parameters using grey-fuzzy-based hybrid techniques. *Materials*, 14(19), 5820.

Sharma, A., Sandhu, H.S., Goyal, D., Goyal, T., Jarial, S., & Sharda, A. (2023a). Sustainable development in cold gas dynamic spray coating process for biomedical applications: Challenges and future perspective review. *International Journal on Interactive Design and Manufacturing (IJIDeM)*, 1–17. https://doi.org/10.1007/s12008-023-01474-7

Sharma, V.K., Rana, M., Singh, T., Singh, A.K., & Chattopadhyay, K. (2021c). Multi-response optimization of process parameters using Desirability Function Analysis during machining of EN31 steel under different machining environments. *Materials Today: Proceedings*, 44, 3121–3126. https://doi.org/10.1016/j.matpr.2021.02.809

Shehata, H.A., Ebeid, S.J., & Mohamed Kohail, A. (2012). Optimization of EDM process parameters for Al-SiC reinforced metal matrix composite. *International Journal of Engineering and Technical Research*, 8(2).

Singh, B.P., Singh, J., Bhayana, M., & Goyal, D. (2021). Experimental investigation of machining nimonic-80A alloy on wire EDM using response surface methodology. *Metal Powder Report*, 76, 9–17. https://doi.org/10.1016/j.mprp.2020.12.001

Singh, B.P., Singh, J., Bhayana, M., Singh, K., & Singh, R. (2022). Experimental examination of the machining characteristics of Nimonic 80-A alloy on wire EDM. *Materials Today: Proceedings*, 69, 291–296. https://doi.org/10.1016/j.matpr.2022.08.537

Singh, J., Singh, C., & Singh, K. (2023). Rotary ultrasonic machining of advance materials: A review. *Materials Today: Proceedings*. https://doi.org/10.1016/j.matpr.2023.01.159

Singh, M., & Maharana, S. (2020). Investigating the EDM parameter effects on aluminium based metal matrix composite for high MRR. *Materials Today: Proceedings*, 33, 3858–3863.

Singh, S., Maheshwari, S., & Pandey, P.C. (2004). Some investigations into the electric discharge machining of hardened tool steel using different electrode materials. *Journal of Materials Processing Technology*, 149(1–3), 272–277.

Singla, M., Dwivedi, D.D., Singh, L., & Chawla, V. (2009). Development of aluminium based silicon carbide particulate metal matrix composite. *Journal of Minerals and Materials Characterization and Engineering*, 8(6), 455.

Srivastava, A. (2020). Assessment of mechanical properties and EDM machinability on Al6063/SiC MMC produced by stir casting. *Materials Today: Proceedings*, 25, 630–634.

Srivastava, A., Dixit, A.R., & Tiwari, S. (2014). A review on fabrication and characterization of aluminium metal matrix composite (AMMC). *International Journal of Advance Research and Innovation*, 2(2), 516–521.

Srivastava, A., Yadav, S.K., & Singh, D.K. (2021). Modeling and optimization of electric discharge machining process parameters in machining of Al 6061/SiCp metal matrix composite. *Materials Today: Proceedings*, 44, 1169–1174.

Talla, G., Sahoo, D.K., Gangopadhyay, S., & Biswas, C.K. (2015). Modeling and multi-objective optimization of powder mixed electric discharge machining process of aluminum/alumina metal matrix composite. *Engineering Science and Technology, an International Journal*, 18(3), 369–373.

Trzepieciński, T., Najm, S.M., & Lemu, H.G. (2022). Current concepts for cutting metal-based and polymer-based composite materials. *Journal of Composites Science*, 6(5), 150.

Uhlmann, E., & Roehner, M. (2008). Investigations on reduction of tool electrode wear in micro-EDM using novel electrode materials. *CIRP Journal of Manufacturing Science and Technology*, 1(2), 92–96.

Verma, D., Gope, P.C., Shandilya, A., Gupta, A., & Maheshwari, M.K. (2013). Coir fibre reinforcement and application in polymer composites. *Journal of Materials and Environmental Science*, 4(2), 263–276.

Wasif, M., Khan, Y.A., Zulqarnain, A., & Iqbal, S.A. (2022). Analysis and optimization of wire electro-discharge machining process parameters for the efficient cutting of Aluminum 5454 alloy. *Alexandria Engineering Journal*, 61(8), 6191–6203.

Yadav, V., Jain, V.K., & Dixit, P.M. (2002). Thermal stresses due to electrical discharge machining. *International Journal of Machine Tools and Manufacture*, 42(8), 877–888.

8 Experimental Investigation on Surface Texture of Inconel-800 with Hybrid Machining Method Using Optimization Technique

Satish Kumar, Harvinder Singh, Rahul Mehra, Aneesh Goyal

8.1 INTRODUCTION

Capacity to bear temperature and surface stability are the unique properties of the Inconel-800 which is very difficult to machine with traditional machining process like EDM. Many researchers have reported the various machining aspects in the conventional EDM, which have some restrictions to accomplish the mirrorlike surface texture. To fulfill today's demand, we need hybrid machining method over conventional machining method. Authors have investigated various methods of EDT for texture and attempted to establish the relationship between surface topography and surface and also applied artificial intelligence techniques for topography prediction. Suhas & Joshi (2018), Lian et al. (2021), and Jain & Parasher (2021) investigated various methods of EDT for texture and attempted to establish the relationship between surface topography and surface and also applied artificial intelligence techniques for topography prediction. Mohanty et al. (2018), Suhas & Joshi (2021), Jeavudeen et al. (2021), Alam et al. (2021), Kumar et al. (2022a), and Goyal et al. (2022) focused on flake problem during processing of titanium alloy with EDM. For this problem, the authors used a B_4C powder additive in spark oil and observed the reduction of recast layer, and this improved the surface quality. Sharma et al. (2021a, 2021b), and Mehra et al. (2020) discussed the effect of process variables on biomedical ingredients, along with their properties which are essential for their biocompatibility. They also reviewed the optimization process on the machining parameters employed for various biomedical applications (Sharma et al., 2021, 2022). PMEDM as a machine and copper as a tool electrode were used on metal matrix composites (MMCs).

DOI: 10.1201/9781003327905-8

Some studies mentioned that when the limits of both additive manufacturing (AM) and subtractive manufacturing (SM) methods are considered, the necessity for hybrid manufacturing as we know it today emerges. The idea transcends the simple post-production machining of additively manufactured components to become a new paradigm with far-reaching implications. The essay discusses the development of the relevant concepts and highlights the primary research and industry difficulties with the help of the relevant scientific literature. In its final analysis, it provides a coherent description of hybrid manufacturing AM/SM and highlights the most important factors in its evolution (Akhai & Rana, 2022; Babbar et al., 2020a, 2020b, 2020c, 2020d, 2021, 2022; Kalia et al., 2022; Khanduja et al., 2021; Kumar et al., 2021; Prakash et al., 2021; Rampal et al., 2021; Rana & Akhai, 2022; Sharma & Jain, 2020; Sharma et al., 2018a, 2018b, 2019a, 2019b, 2020a, 2020b, 2020c, 2022c, 2023a, 2023b, 2023c; Sharma V K et al., 2021; Singh et al., 2021, 2022a, 2023).

Various techniques have been used for the optimization of the result and its validation. Kumar et al. (2021) examined surface characteristics of titanium alloys with powdered EDM using carborundum abrasive powder. They investigated the significant effect on MRR and surface finish, with different process parameters. Chauhan et al. (2022) and Kumar et al. (2021) investigated the stability and uniformity of powders in terms of their particle size and concentration in the powder mixed dielectric. The outcomes showed that through TOPSIS analysis, proximity value improved by 2.39% compared to the expected value. Singh et al.'s (2021) machining of titanium alloys with powder mixed EDM has been investigated with graphite and titanium oxide as the powder and kerosene as the dielectric. MRR, TWR, and surface roughness have been selected for the response parameters. It is evident from the previous iterative review that work done by researchers are not sufficient on the PMEDM of the Inconel-800. Therefore, we need to develop a method in which powder is added to the dielectric oil to enhance the surface properties.

8.2 MATERIALS AND METHODS

For the selection of final process parameters, pilot experiments were performed with metal powder mentioned in the Table 8.3 changing from 1 to 12 g/l in the EDM oil on the die-sinking EDM machine as shown in Figure 8.1 (a) and (b). Better surface roughness achieved with powder concentration of 7 g/l. Therefore, for final experimentation powder concentration 7 g/l has been taken.

For further experimental study, Cu, Cu-Cr, and Gr were chosen as the electrodes. The chemical composition of the work samples and electrode materials are presented in Table 8.1 and 8.2.

Figure 8.2 shows the workpiece in the form of rectangular shape. Table 8.3 reveals the process variables along with their levels. The Box-Benken design of RSM was used for the experiments.

8.2.1 RSM

It discovered the connection between different input variables and response parameters. By creating contour plots and residual plots to check for accuracy, it examines the data. The equation for RSM is the following:

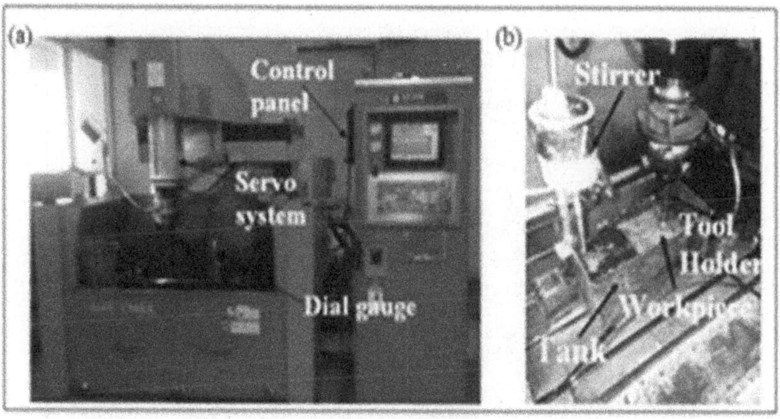

FIGURE 8.1 (a), (b) Schematic diagram and experimental setup for EDM machine of Inconel-800.

TABLE 8.1
Material Inconel-800's Chemical Composition

Element	Ni	Fe	Cr	C	Mn	S	Si	Cu
%	Base material	7.78	14.56	0.15	1	0.015	0.5	0.5

TABLE 8.2
Chemical Evaluation of Electrode Materials (wt%)

Elements	Copper	Copper-Chromium
Cu	99.1	98.4
Zn	0.0148	<0.0050
Pb	0.0206	0.0118
Sn	0.0356	<0.0050
Mn	0.005	0.006
Fe	0.109	0.0319
Ni	0.0083	0.0104
Si	<0.0050	<0.0050
Cr	0.0061	1.36
Al	<0.0020	<0.0020
S	<0.0020	<0.0020
Bi	<0.0050	<0.0050
Sb	<0.0050	0.0072

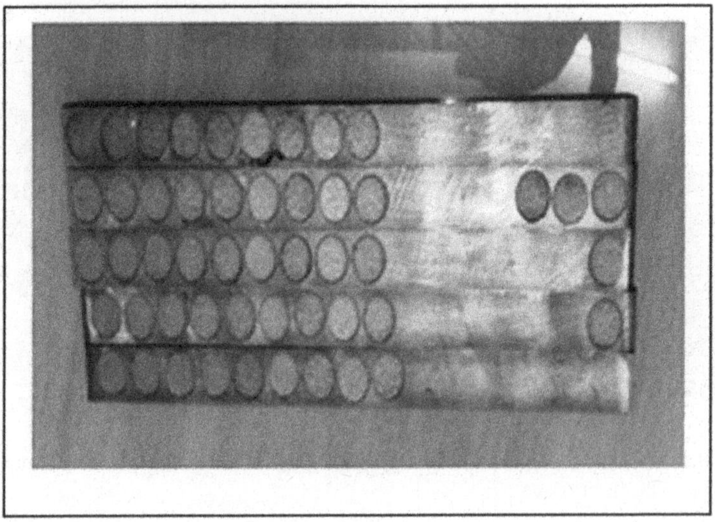

FIGURE 8.2 Workpiece material for Inconel-800 after machining.

TABLE 8.3
Variables in the Machining Process and Their Levels

S. No	Symbols	Input Factors	Level I	Level II	Level III	Units
1	A	Current (Ip)	4	5	12	Ampere
2	B	Pulse-on time (T_{on})	60	90	120	µs
3	C	Pulse-off time (T_{off})	30	45	60	µs
4	D	Tool material	Cu	Cu-Cr	Graphite	Ø 12mm
5	E	Powder particles	WC	Cobalt	B_4C	gram

$$y_u = \alpha_0 + \sum_{i=1}^{L} \alpha_i x_i + \sum_{i=1}^{L} \alpha_{ii} x^2{}_i + \sum_{i<1} \alpha_{ij} x_i x_j \mathrm{K\ K\ K}, \qquad (1)$$

Where xi (1, 2, . . . , L) are coded levels of L process variables, Yu is the response, and the terms are the IInd order regression coefficients. For this, Eq. (1) can be modified according to the five variables used as follows:

$$\begin{aligned} y_u = {} & \alpha_0 + \alpha_0 x_1 + \alpha_2 x_2 + \alpha_3 x_3 + \alpha_4 x_4 + \alpha_5 x_5 + \alpha_{11} x_{12} + \alpha_{22} x_{22} + \alpha_{33} x_{32} \\ & + \alpha_{44} x_{42} + \alpha_{55} x_{52} + \alpha_{12} x_1 x_2 + \alpha_{13} x_1 x_3 + \alpha_{14} x_1 x_4 + \alpha_{15} x_1 x_5 + \alpha_{23} x_2 x_3 \\ & + \alpha_{24} x_2 x_4 + \alpha_{25} x_3 x_5 \mathrm{K\ K\ K}, \end{aligned} \qquad (2)$$

8.3 RESULTS AND DISCUSSIONS

8.3.1 INVESTIGATION OF SR

Table 8.4 shows the combination of input variables and output (i.e. SR).

From Figure 8.3 (a), it is revealed that when current is increased, SR also hikes. When both T_{off} and T_{on} hike, SR has a fraction effect. Figure 8.3 (b) and (c) illustrate

TABLE 8.4
Design of Experiments and Results

	Input Factors					Response			
Run No.	Current (Amp)	Pulse-on time (T_{on}) (μs)	Pulse-off time (T_{off}) (μs)	Tool Material	Powder	Surface Roughness (Ra)			Mean Value
	(A)	(B)	(C)	(D)	(E)	R-1	R-2	R-3	
1	4	90	60	Cu-Cr	Co	4.503	4.504	4.511	4.506
2	8	90	45	Cu-Cr	Co	7.918	7.92	7.952	7.93
3	8	90	60	Gr	BC	6.45	6.482	6.466	6.466
4	8	90	45	Cu	WC	9.89	9.892	9.876	9.886
5	8	90	30	Cu-Cr	BC	7.156	7.206	7.406	7.256
6	8	120	30	Cu-Cr	Co	9.16	8.95	9.67	9.26
7	8	90	60	Cu-Cr	BC	5.9	6.11	6.65	6.22
8	8	90	60	Cu	Co	8.88	8.78	9.1	8.92
9	4	90	45	Cu	Co	4.01	4.81	4.41	4.41
10	8	90	45	Cu-Cr	Co	7.718	7.72	7.752	7.73
11	8	90	45	Cu-Cr	Co	7.7	7.76	7.82	7.76
12	8	90	45	Gr	WC	8.333	8.38	8.616	8.443
13	8	90	30	Gr	Co	9.68	9.699	9.769	9.716
14	12	90	30	Cu-Cr	Co	11.715	11.698	10.256	11.223
15	8	90	30	Cu	Co	6.15	6.1	6.2	6.15
16	12	90	45	Cu-Cr	BC	9.175	9.187	9.217	9.193
17	12	90	60	Cu-Cr	Co	6.435	6.475	8.096	7.002
18	8	120	45	Cu-Cr	WC	8.56	8.56	8.569	8.569
19	4	90	45	Gr	Co	5.1	5.1	5.28	5.16
20	12	90	45	Gr	Co	10.58	10.602	10.627	10.603
21	8	120	60	Cu-Cr	Co	7.11	7.1	7.15	7.12
22	4	90	45	Cu-Cr	BC	3.702	3.701	3.706	3.703
23	8	90	45	Gr	BC	8.88	8.882	8.887	8.883
24	8	120	45	Cu	Co	9.45	8.42	7.63	8.5

(Continued)

TABLE 8.4 *(Continued)*
Design of Experiments and Results

	Input Factors					Response			
Run No.	Current (Amp)	Pulse-on time (T_{on}) (μs)	Pulse-off time (T_{off}) (μs)	Tool Material	Powder	Surface Roughness (Ra)			Mean Value
	(A)	(B)	(C)	(D)	(E)	R-1	R-2	R-3	
25	12	60	45	Cu-Cr	Co	9.55	10.5	11.66	10.57
26	8	90	45	Cu-Cr	C	7.33	7.32	7.34	7.33
27	8	90	60	Cu-Cr	WC	7.84	7.84	7.87	7.85
28	4	90	30	Cu-Cr	Co	3.67	3.67	3.679	3.673
29	12	90	45	Cu	Co	9.9	9.7	9.8	9.8
30	4	120	45	Cu-Cr	Co	4.56	4.562	4.567	4.563
31	8	120	60	Cu-Cr	Co	6.6	6.61	6.65	6.62
32	12	90	45	Cu-Cr	WC	8.984	8.983	8.991	8.986
33	8	90	45	Cu	BC	5.505	5.505	5.508	5.506
34	8	90	45	Cu-Cr	Co	8.3	8.31	8.38	8.33
35	8	60	45	Cu-Cr	WC	7.031	7.032	7.036	7.033
36	12	120	45	Cu-Cr	Co	9.092	9.091	9.096	9.093
37	8	120	45	Gr	Co	8.025	8.025	8.028	8.026
38	8	120	45	Cu-Cr	BC	7.4	7.35	7.6	7.45
39	8	90	45	Cu-Cr	Co	7.71	7.707	7.812	7.743
40	8	90	30	Cu-Cr	WC	9.28	9.275	9.234	9.263
41	8	60	60	Cu-Cr	Co	7.88	7.81	7.998	7.896
42	4	60	45	Cu-Cr	Co	2.801	8.795	2.735	2.777
43	8	60	45	Cu-Cr	BC	7.443	7.493	7.693	7.543
44	8	60	45	Gr	Co	8.5	8.6	9.09	8.73
45	8	60	45	Cu	Co	7.73	7.69	6.348	7.256
46	4	90	45	Cu-Cr	WC	4.001	4.009	4.029	4.013

this, while SR increased when graphite is used as a tool electrode. Figure 8.3 (e) shows that B_4C has the vital effect on SR.

Ideal parameters for SR during PMEDM of Inconel-800 are Ip = 4 A, T_{on} = 60 s, T_{off} = 60 s, Cu-Cr as an electrode, and B4C as a powder. The perturbation plot is displayed in Figure 8.3 (f) in order to illustrate the overall impact of all the factors. After backward elimination, ANOVA for SR is shown in Table 8.5. Non-vital terms are removed by using backward elimination process so that SR is fit using a quadratic model. According on ANOVA shown in Table 8.5, terms except B, A2, D2, and AB are vital terms.

$$\text{Eq. for SR} = +8.7475 + 3.57 \times A + 1.36 \times B - 1.57 \times C + 1.38 \times D - 1.056 \times E - 2.11 \times A2 + 0.43\times D2 - 1.53 \times A \times C - 1.99 \times B \times C - 1.75 \times B \times D - 1.67 \times B \times E - 1.44 \times C \times D$$

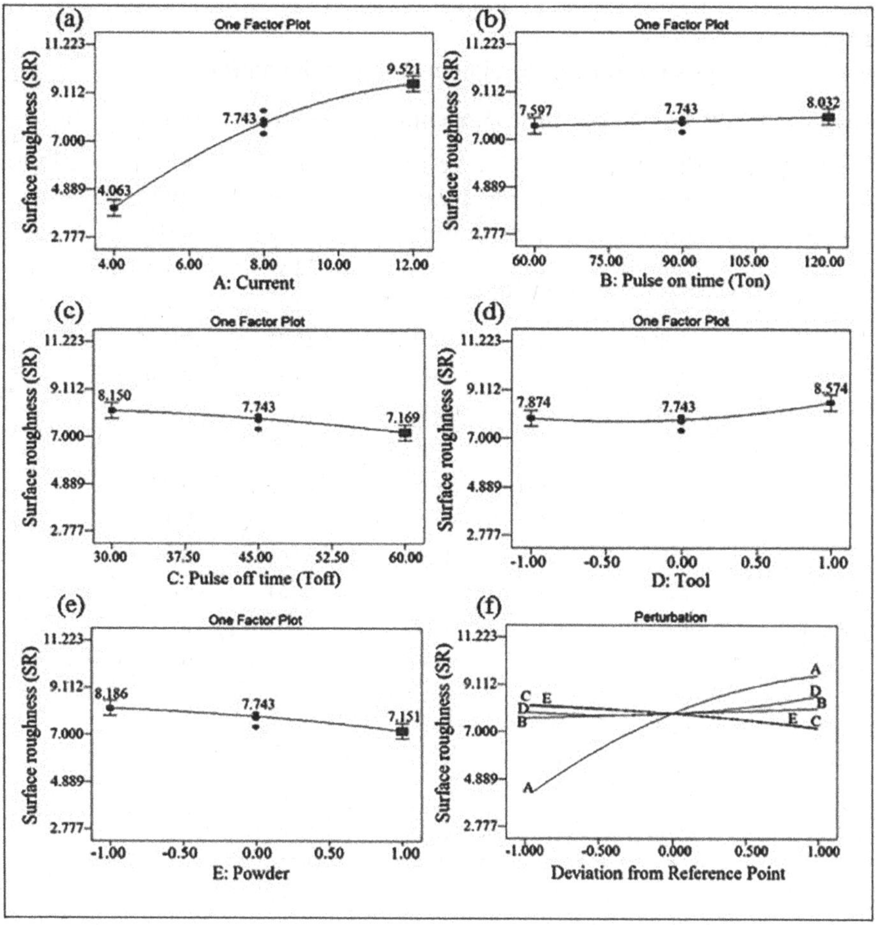

FIGURE 8.3 Surface roughness displayed as one factor.

The value of R^2 is .9503, Pred-R^2 = .8911, and Adj- R^2 = .9322 along with high correlation between the "Pred R^2" of.8911 and the "Adj R^2" of.9322. Arriving with a p-value of 0.1003 means this model is fit for data. For factors A and E, counter plot is shown in Figure 8.4 (a), whereas normal plots of residuals are shown in Figure 8.4 (d). From the figure, it is observed that all the outcomes gotten are near to the predicted terms. 3D plots and Figure 8.4 (b), (c) also showed that T_{on} has very fraction influence on SR.

8.4 SEM MICROGRAPH ANALYSIS

Surface texture was examined by SEM, while the machined surface superiority is shown by the crack creation, pits, and debris on the sample surface. All parameters we took for study are the main factors that are responsible for machined surface superiority (Kumar & Kumar, 2022; Kumar & Singh, 2022; Kumar &

TABLE 8.5
Variance Analysis for Surface Roughness (After Elimination)

Source	Sum of Squares	DF	Mean Square	F Value	Prob>F	
Model	169.98	12	14.17	52.54	<0.0001	significant
A	119.16	1	119.16	442.03	<0.0001	
B	0.69	1	0.69	2.56	0.1189	
C	3.55	1	3.55	13.17	0.0009	
D	1.96	1	1.96	7.27	0.011	
E	4.29	1	4.29	15.91	0.0003	
A2	9.4	1	9.4	34.88	<0.0001	
D2	2.21	1	2.21	8.18	0.0073	
AB	2.66	1	2.66	9.87	0.0035	
AC	6.39	1	6.39	23.69	<0.0001	
BC	1.79	1	1.79	6.64	0.0146	
CD	9.06	1	9.06	33.61	<0.0001	
DE	5.81	1	5.81	21.54	<0.0001	
Residual	8.9	33	0.27			
Lack of Fit	8.24	27	0.31	2.8	0.1003	not significant
Pure Error	0.65	6	0.11			
Cor Total	178.88	45				
R^2	0.9503		Adj R2	0.9322		
Pred R^2	0.8911		Adeq Pre	29.873		

Dhingra, 2015, 2018). Magnification power of SEM images are 1.00 X and 3.00 KX respectively. Micrograph of Inconel-800 machined with Cu electrode in B_4C powder mixed dielectric fluid at 8A current, 90 μs T_{on}, and 60 μs T_{off} is shown in Figure 8.5 (a) and (d). Figure 8.5 (a) describes the creation of micro cracks on the texture, but with advanced magnification, sub-cracks can be observed on the texture shown in Figure 8.5 (d). Figure 8.5 (b) and (e) showed the micro holes and hole on the surface, but there are no cracks noticed with Ip 4A, T_{on} corresponds to the 90 μs, while T_{off} is now 30 μs. SR is reduced with low current as shown in Figure 8.5 (b), but few sub-cracks were noticed with higher magnification as shown in Figure 8.5 (e). With machining condition, 12 A current, 90 μs T_{on}, and 45μs T_{off}, micrograph is shown in Figure 8.5(f). During the serious examination of the specimen, surface defects mentioned earlier also melted carbon particles

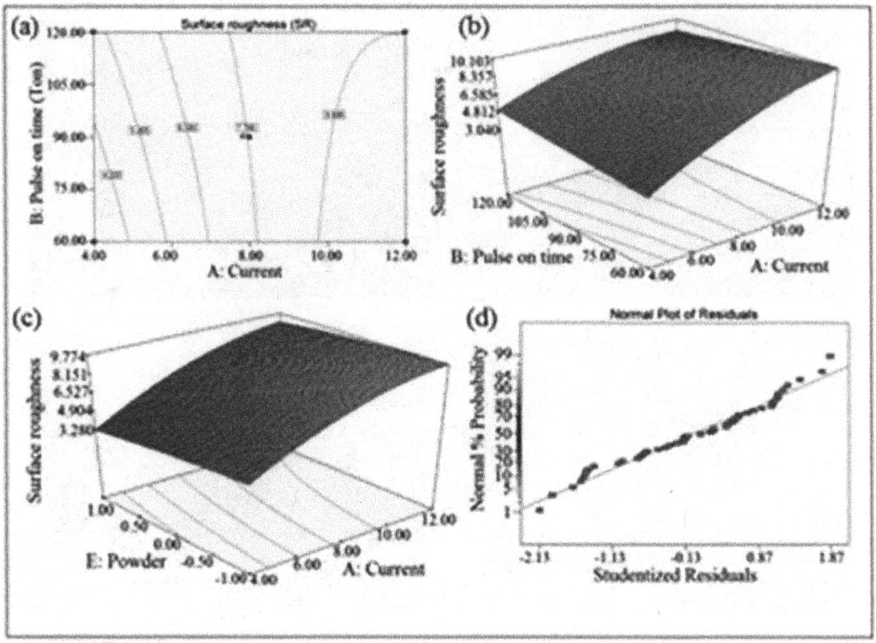

FIGURE 8.4 Surface plots for the response: (a) SR counter plot, (b) and (c) 3D surface plots, and (d) visualization of normal probability residuals.

and craters of different sizes, and surface cracks observed resulted in a poor surface finish. This is due to the reason that as the current hikes, specimen melts with the heat energy.

8.5 CONCLUSION

Based on investigational study, the SR variation is between the 4.063 mm and 9.521 mm. Except T_{on}, all other vital factors affected the SR. It improves with T_{off} and Cu-Cr as an electrode, and B_4C is taken as powder elements while it hiked with the hike of Ip. T_{on} has a fraction effect on SR. R^2, adj. R^2, and predicted R^2 found for SR is 0.9503, 0.9322, and 0.8911. The "Pred R^2" of .8911 and the "Adj R^2" of 0.9322 concur satisfactorily. The surface morphology of the material is negatively impacted by peak current, resulting in the development of fractures, holes, and debris, but improved results may be achieved with reduced current and greater T_{off}, according to SEM micrographs.

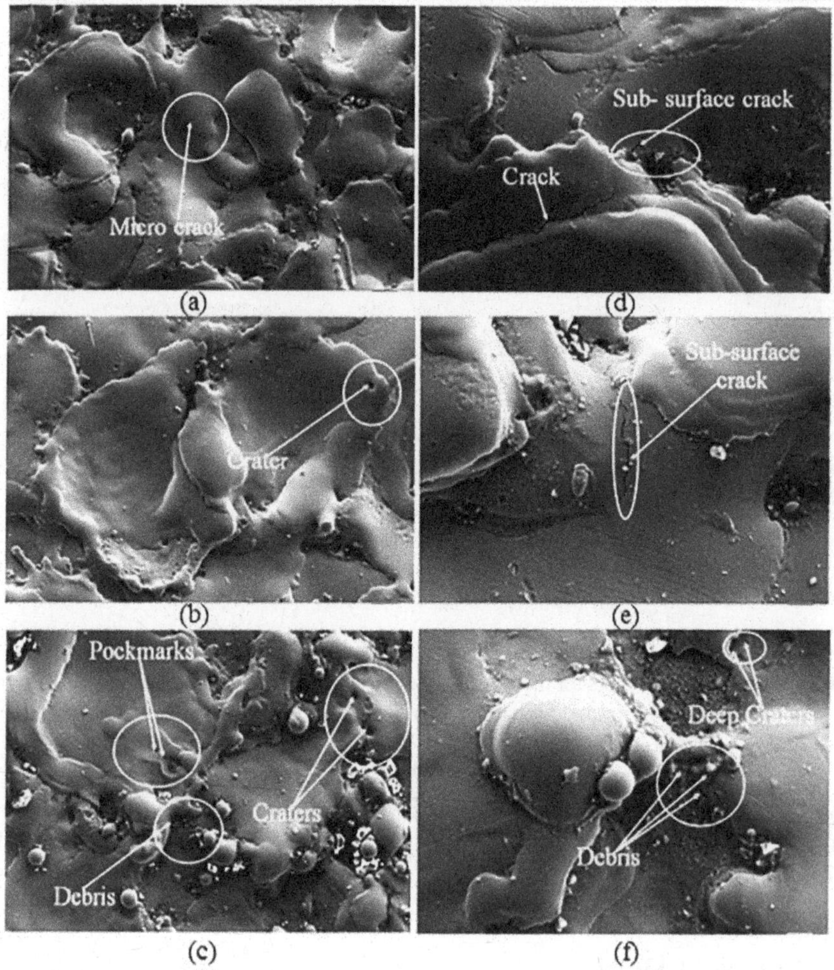

FIGURE 8.5 Shows the SEM images (500* and 2.000 KX) of experiment no. 8 at Ip = 8 A, T_{on} = 90 s, and T_{off} = 60 s, using a Cu electrode in a blended dielectric of B_4C powder; at Ip = 8 A, T_{on} = 90 s, T_{off} = 30 s, and Cu-Cr tool in blended WC powder dielectric, trial number 13 (b and e); and trial no. 16 (c and f), at Ip = 12 A, T_{on} = 90 s, T_{off} = 45 s, dielectric using a Cu tool and B_4C powder.

REFERENCES

Akhai, S., & Rana, M. (2022). Taguchi-based grey relational analysis of abrasive water jet machining of Al-6061. *Materials Today: Proceedings*, 65, 3165–3169. https://doi.org/10.1016/j.matpr.2022.05.361

Alam, S.T., Amin, A.K.N., Hossain, I., Huq, M., & Tamim, S.H. (2021). Performance evaluation of graphite and titanium oxide powder mixed dielectric for electric discharge machining of Ti—6Al—4V. *SN Applied Sciences, A Springer Nature Journal*, 3, 435.

Babbar, A., Jain, V., Gupta, D., Prakash, C., & Sharma, A. (2020a). Fabrication and machining methods of composites for aerospace applications. In *Characterization, Testing, Measurement, and Metrology* (1st ed., pp. 109–124). Boca Raton, FL: CRC Press.

Babbar, A., Jain, V., Gupta, D., Prakash, C., Singh, S., & Sharma, A. (2020b). Effect of process parameters on cutting forces and osteonecrosis for orthopedic bone drilling applications. In *Characterization, Testing, Measurement, and Metrology* (1st ed., pp. 93–108). Boca Raton, FL: CRC Press.

Babbar, A., Jain, V., Gupta, D., Prakash, C., Singh, S., & Sharma, A. (2020d). 3D bioprinting in pharmaceuticals, medicine, and tissue engineering applications. In *Advanced Manufacturing and Processing Technology* (1st ed., pp. 147–161). Boca Raton, FL: CRC Press, 2021.

Babbar, A., Jain, V., Gupta, D., & Sharma, A. (2020c). Fabrication of microchannels using conventional and hybrid machining processes. In *Non-Conventional Hybrid Machining Processes* (1st ed., pp. 37–51). Boca Raton, FL: CRC Press.

Babbar, A., Rai, A., & Sharma, A. (2021). Latest trend in building construction: Three-dimensional printing. *Journal of Physics: Conference Series*, 1950, 012007.

Babbar, A., Sharma, A., & Singh, P. (2022). Multi-objective optimization of magnetic abrasive finishing using grey relational analysis. *Materials Today: Proceedings*, 50, 570–575.

Chauhan, A., Kumar, M., & Kumar, S. (2022). Fabrication of polymer hybrid composites for automobile leaf spring application. *Material Today Proceeding*, 48(5), 1371–1377. https://doi.org/10.1016/j.matpr.2021.09.114

Goyal, A., Singh, H., Goyal, R., Singh, R., & Singh, S. (2022). Recent advancements in abrasive flow machining and abrasive materials: A review. *Materials Today: Proceedings.* https://doi.org/10.1016/j.matpr.2021.12.109

Jain, S., & Parasher, V. (2021). Critical review on the impact of EDM process on biomedical materials. *Materials and Manufacturing Processes*, 36(15), 1701–1724.

Jeavudeen, S., Zailani, H.S., & Murugan, M. (2021). Enhancement of machinability of titanium alloy in the Eductor based PMEDM process. *SN Applied Sciences, A Springer Nature Journal*, 3, 490.

Kalia, G., Sharma, A., & Babbar, A. (2022). Use of three-dimensional printing techniques for developing biodegradable applications: A review investigation. *Materials Today: Proceedings*, 62, 346–352.

Khanduja, P., Bhargave, H., Babbar, A., Pundir, P., & Sharma, A. (2021). Development of two-dimensional plotter using programmable logic controller and human machine interface. *Journal of Physics: Conference Series*, 1950, 012012.

Kumar, R., Kumar, M., Chauhan, J.S., & Kumar, S. (2022a). Overview on metamaterial: History, types and applications. *Material Today Proceedings*, 56(5), 3016–3024. https://doi.org/10.1016/j.matpr.2021.11.423

Kumar, R., & Kumar, S. (2022). Conventional and 3D Printing technology for the manufacturing of metal matrix composite: A study. In *Metal Matrix Composites, Fabrication, Production and 3D Printing.* Boca Raton, FL: CRC Press, eBook.

Kumar, S., & Dhingra, A.K. (2015). Multiresponse optimization of process variables of powder mixed electrical discharge machining on inconel-600 using taguchi methodology. *International Journal of Mechanical and Production Engineering Research and Development (IJMPERD) Journal Scopus Indexed Journal*, 60(6).

Kumar, S., & Dhingra, A.K. (2018). Effect of machining parameters on performance characteristics of powder mixed EDM of inconel-800. *International Journal of Automotive and Mechanical Engineering*, 15(2), 5221–5237. Scopus Indexed Journal, https://doi.org/10.15282/ijame.15.2.2018.6.0403, ISSN: 2229–8649 (Print); ISSN: 2180–1606 (Online).

Kumar, S., Kumar, S., Sharma, R., Bishnoi, P., Singh, M., & Singh, R. (2022b). To evaluate the effect of boron carbide (B4C) powder mixed EDM on the machining characteristics of INCONEL-600. *Materials Today: Proceedings*, 56, 2794–2799. https://doi.org/10.1016/j.matpr.2021.10.096

Kumar, S., Kumar, S., Singh, R., Bishnoi, P., & Chahal, V. (2021). Analysis of PMEDM parameters for the machining of inconel-800 material using Taguchi methodology, advances in materials and mechanical engineering. In *Advances in Materials and Mechanical Engineering: Select Proceedings of ICFTMME 2020* (pp. 321–328). Springer Singapore.

Kumar, S., & Singh, S. (2022). Corrosion behavior of metal, alloy and composite: An overview. In *Metal Matrix Composites: Properties and Application*. CRC Press, eBook ISBN9781003194910. www.taylorfrancis.com/books/edit/10.1201/9781003194910/metal-matrix-composites-suneev-anil-bansal-virat-khanna-pallav-gupta

Kumar, S., Sudhakar, R.P., Goyal, D., & Sehgal, S. (2021). Process modelling for machining Inconel 825 using cryogenically treated carbide insert. *Metal Powder Report*, 76, 66–74. https://doi.org/10.1016/j.mprp.2020.06.001

Lian, M.G., Chen, S.G., Lei, J.G., Wu, X.Y., Guo, C., Peng, T.J., Yang, J., Luo, F.L., & Zhao, H. (2021). Combining PMEDM with the tool electrode sloshing to reduce recast layer of titanium alloy generated from EDM. *The International Journal of Advanced Manufacturing Technology*, 11, 1535–1545.

Mehra, R., Mohal, S., Lonia, B., & Kumar, M. (2020). Optimization of laser engraving process parameters for the engraving of hybrid glass fiber reinforced plastic (GFRP) combinations. *Laser in Engineering, LIE*, 45, 4–6.

Mohanty, S., Mishra, A., Nanda, B.K., & Routara, B.C. (2018). Multi-objective parametric optimization of nano powder mixed electrical discharge machining of AlSiCp using response surface methodology and particle swarm optimization. *Alexandria Engineering Journal*, 57(2), 609–619.

Prakash, C., Kumar, V., Mistri, A., Sharma, A., Uppal, A.S., Babbar, A., & Pathri, B.P. (2021). Investigation of functionally graded adherents on failure of socket joint of FRP composite tubes. *Materials*, 14, 6365.

Rampal, R., Goyal, T., Goyal, D., Mittal, M., Dang, R.K., & Bahl, S. (2021). Magneto-rheological abrasive finishing (MAF) of soft material using abrasives. *Materials Today: Proceedings*, 45, 51140–5121. https://doi.org/10.1016/j.matpr.2021.01.629

Rana, M., & Akhai, S. (2022). Multi-objective optimization of Abrasive water jet Machining parameters for Inconel 625 alloy using TGRA. *Materials Today: Proceedings*, 65, 3205–3210. https://doi.org/10.1016/j.matpr.2022.05.374

Sharma, A., Babbar, A., Jain, V., & Gupta, D. (2018b). Enhancement of surface roughness for brittle material during rotary ultrasonic machining. *MATEC Web of Conferences*, 249, 01006.

Sharma, A., Babbar, A., Tian, Y., Pathri, B.P., Gupta, M., & Singh, R. (2022a). Machining of ceramic materials: A state of the art review. *International Journal on Interactive Design and Manufacturing (IJIDeM)*, 1–21. https://doi.org/10.1007/s12008-022-01016-7

Sharma, A., Fidan, I., Huseynov, O., Ali, M.A., Alkunte, S., Rajeshirke, M., Gupta, A., Hasanov, S., Tantawi, K., Yasa, E., Yilmaz, O., Loy, J., & Popov, V. (2023b). Recent inventions in additive manufacturing: Holistic review. *Inventions*, 8(4), 103. https://doi.org/10.3390/inventions8040103

Sharma, A., Grover, V., Babbar, A., & Rani, R. (2020c). A trending nonconventional hybrid finishing/machining process. In *Non-Conventional Hybrid Machining Processes* (1st ed., pp. 79–93). Boca Raton, FL: CRC Press.

Sharma, A., & Jain, V. (2020). Experimental investigation of cutting temperatureduring drilling of float glass specimen. In IOP Conference Series: Materials Science and Engineering (Vol. 715, No. 1, p. 012050). IOP Publishing. https://doi.org/10.1088/1757-899X/715/d1/012050

Sharma, A., Jain, V., & Gupta, D. (2018a). Characterization of chipping and tool wear during drilling of float glass using rotary ultrasonic machining. *Measurement*, 128, 254–263.

Sharma, A., Jain, V., & Gupta, D. (2019a). Comparative analysis of chipping mechanics of float glass during rotary ultrasonic drilling and conventional drilling: For multi-shaped tools. *Machining Science and Technology*, 23, 547–568.

Sharma, A., Jain, V., & Gupta, D. (2019b). Multi-shaped tool wear study during rotary ultrasonic drilling and conventional drilling for amorphous solid. *Proceedings of the Institution of Mechanical Engineers, Part E: Journal of Process Mechanical Engineering*, 233, 551–560.

Sharma, A., Jain, V., & Gupta, D. (2021a). Effect of pre and post tempering on hole quality of float glass specimen: For rotary ultrasonic and conventional drilling. *Silicon*, 13, 2029–2039.

Sharma, A., Jain, V., & Gupta, D. (2022b). Mathematical approach on chipping volume estimation generated during rotary ultrasonic drilling for float glass. *Proceedings of the National Academy of Sciences, India Section A: Physical Sciences*, 92, 285–291.

Sharma, A., Jain, V., Gupta, D., & Babbar, A. (2020a). A review study on miniaturization. In *Advanced Manufacturing and Processing Technology* (1st ed., pp. 111–131). Boca Raton, FL: CRC Press.

Sharma, A., Jain, V., Gupta, D., & Babbar, A. (2020b). A review study on miniaturization. In *Advanced Manufacturing and Processing Technology* (1st ed., pp. 111–131). Boca Raton, FL: CRC Press, 2021.

Sharma, A., Kalsia, M., Uppal, A.S., Babbar, A., & Dhawan, V. (2022c). Machining of hard and brittle materials: A comprehensive review. *Materials Today: Proceedings*, 50, 1048–1052.

Sharma, A., Kumar, V., Babbar, A., Dhawan, V., & Kotecha, K. (2021b). Experimental investigation and optimization of electric discharge machining process parameters using grey-fuzzy-based hybrid techniques. *Materials*, 14(19), 5820.

Sharma, A., Parikh, P.A., Roy, D., Joshi, K., & Trivedi, R. (2023c). Performance evaluation of an indigenously-designed high-performance dynamic feeding robotic structure using advanced additive manufacturing technology, machine learning and robot kinematics. *International Journal on Interactive Design and Manufacturing (IJIDeM), 1–29.*

Sharma, A., Sandhu, H.S., Goyal, D., Goyal, T., Jarial, S., & Sharda, A. (2023a). Sustainable development in cold gas dynamic spray coating process for biomedical applications: Challenges and future perspective review. *International Journal on Interactive Design and Manufacturing (IJIDeM)*, 1–17. https://doi.org/10.1007/s12008-023-01474-7

Sharma, V.K., Rana, M., Singh, T., Singh, A.K., & Chattopadhyay, K. (2021c). Multi-response optimization of process parameters using Desirability Function Analysis during machining of EN31 steel under different machining environments. *Materials Today: Proceedings*, 44, 3121–3126. https://doi.org/10.1016/j.matpr.2021.02.809

Singh, B.P., Singh, J., Bhayana, M., & Goyal, D. (2021). Experimental investigation of machining nimonic-80A alloy on wire EDM using response surface methodology. *Metal Powder Report*, 76, 9–17. https://doi.org/10.1016/j.mprp.2020.12.001

Singh, B.P., Singh, J., Bhayana, M., Singh, K., & Singh, R. (2022a). Experimental examination of the machining characteristics of Nimonic 80-A alloy on wire EDM. *Materials Today: Proceedings*, 69, 291–296. https://doi.org/10.1016/j.matpr.2022.08.537

Singh, G., Singh, M., Singh, R., Mohal, S., Kumar, S., & Kumar, S. (2022c). Numerical approach for solution of fluid and heat transfer coupled problem through porous media. *Materials Today: Proceedings*, 56, 3031–3034. https://doi.org/10.1016/j.matpr.2021.11.631

Singh, H., Kumar, M., & Singh, R. (2022b). An overview of various applications of cold spray coating process. *Materials Today: Proceedings*. https://doi.org/10.1016/j.matpr.2021.10.160

Singh, J., Singh, C., & Singh, K. (2023). Rotary ultrasonic machining of advance materials: A review. *Materials Today: Proceedings*. https://doi.org/10.1016/j.matpr.2023.01.159

Suhas, S.J., & Joshi, S. (2021). Surface topography generation and simulation in electrical discharge texturing: A review. *Journal of Materials Processing Technology*, 298, 117–297. ISBN9781003194897. www.taylorfrancis.com/books/edit/10.1201/9781003194897/metal-matrix-composites-suneev-anil-bansal-virat-khanna-pallav-gupta

9 Advanced Finishing Processes for Cylindrical Surface Finishing

A Review

Manpreet Singh, Gagandeep Singh, Mohammad Alshinwan

9.1 INTRODUCTION

The term "roughness" is used to describe the extent to which a surface deviates from being perfectly flat in a normal direction. While it is possible to make a surface as smooth as possible, the roughness value will never be able to be decreased to zero. Surface finishing is the process of smoothing out a product's exterior. Surface finishing of components is currently a crucial industry requirement for their precise operational functionality (Balogun & Mativenga, 2017). Some industrial products require surface finishing just for having the good appearance or aesthetic look. However, many industrial products need a finished surface with good surface characteristics in order to achieve benefits, such as dimensional precision, near resistance fit or shape, improved tool life, and lower wear and frictional losses, among other things (Mohanty et al., 2013). Various industries like electronics, optics, automotive, and avionics require specific surface characteristics of micro products and components for their precise operational requirements (Rjurkar et al., 2006). However, it can be challenging to give industrial components' surfaces the necessary precise surface finish. In the manufacturing industry, finishing components that need higher precision is a complex and time-consuming job (Pattnaik et al., 2012). Almost 10% to 15% of the total production cost of a component is expended for its precise surface finishing (Verma et al., 2017).

9.2 CYLINDRICAL SURFACE FINISHING PROCESSES

9.2.1 Conventional Finishing Processes

The surface polish of the individual components can be achieved by a number of traditional finishing procedures used in industry, including honing, lapping, and grinding. But these traditional finishing operations might sometime provide the finished surface with some defects, such as cracks, residual stresses, etc. (Benardos &

DOI: 10.1201/9781003327905-9

Vosniakos, 2003). Various conventional finishing processes are explained as under.

9.2.2 Lapping

Lapping is the oldest finishing operation which is commonly used for finishing of flat surfaces. In spite of its slow pace and low pressure, the resulting structure is remarkably well-fitted and accurate (Khanna & Lal, 2010).

Three-body abrasive wear is the basic operating basis (Chang et al., 2000; Jha & Jain, 2005). It employs a slurry of loose abrasive materials, such as silicon carbide, aluminium oxide, etc. When using a lap plate as a backup wheel, loose abrasives are sandwiched between the workpiece and the wheel. When abrasives are pressed on the surface of a workpiece, the surface is smoothed out. Flat workpieces are finished using a lapping technique, the schematic of which is seen in Figure 9.1.

9.2.3 Honing

Honing is the most common finishing technique for completing the inner surface of a cylindrical workpiece (Pawlus et al., 2009). It is a procedure that is typically used during a boring or drilling operation to achieve a smooth finish over the internal surface of application (Sabri et al., 2009). Figure 9.2 shows the honing method, as it finishes the internal surface of work-part. It consists of various honing stones having abrasives embedded in it. Within the cylindrical parts, the honing instrument is rotated and reciprocated at the same time. The roughness crests found on the inner surface of the workpiece are slashed out, and the finished surface is obtained due to the action (rotation and reciprocation) of the embedded abrasives. The process is used for the components which require precise geometric and form accuracy (Dimkovski

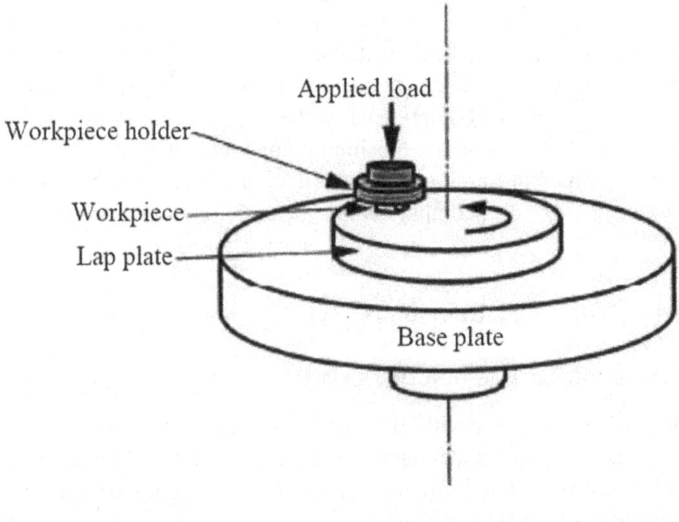

FIGURE 9.1 Schematic of the lapping process performs the finishing on the flat workpiece.

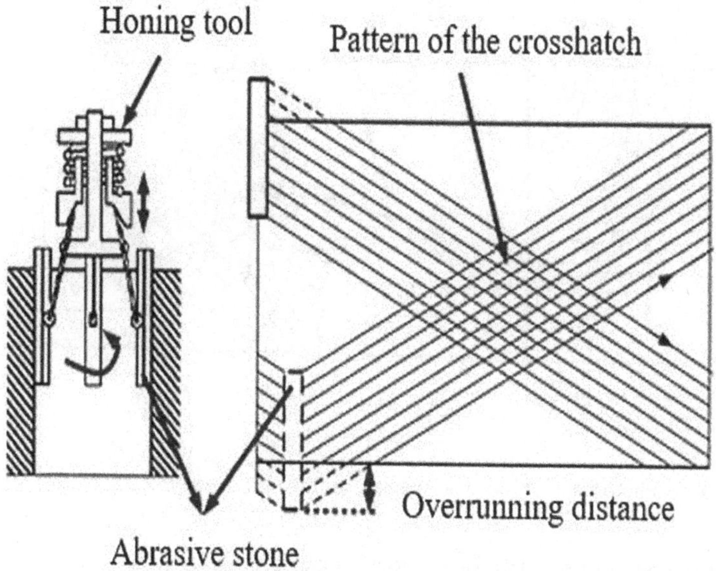

FIGURE 9.2 Schematic of honing process while it performs finishing.

et al., 2009; Sabri et al., 2011; Spencer et al., 2011; Shaji & Radhakrishnan, 2003). It is also helpful in manufacturing the components where good dimensional tolerance is required.

9.2.4 Grinding

For a variety of purposes, including the final polishing of surfaces, grinding is a common technique. It's versatile enough to work with anything from a soft material to something extremely hard (Beyerer & Leon, 1997). Figure 9.3 shows a picture of the experimental grinding setup. The grinding process only uses a circular grinding wheel with silicon carbide (SiC) abrasives embedded around the outside of the wheel. The grinding wheel is rotated over the surface to be finished.

The relative movement between the workpiece surface and the abrasives removes the material in the form of microchips. The previously stated traditional finishing processes have various limitations due to which these processes cannot be opted for manufacturing the very high precise industrial components. Conventional finishing processes are unable to finish complicated surfaces which need a very detailed surface finish at the nanometer level. Since these methods use rigid tools for finishing, they have no control over the finishing forces. As a consequence, these processes may also result in degraded surfaces or sub-surfaces (Lawrence & Ramamoorthy, 2011). Due to the high amount of heat generated during the surface finishing process in conventional finishing systems, problems such as surface crack forming and residual stresses have been observed (Gupte et al., 2008; Jain, 2009). Various innovative finishing methods have flourished to improve surface finish consistency in order to address these limitations.

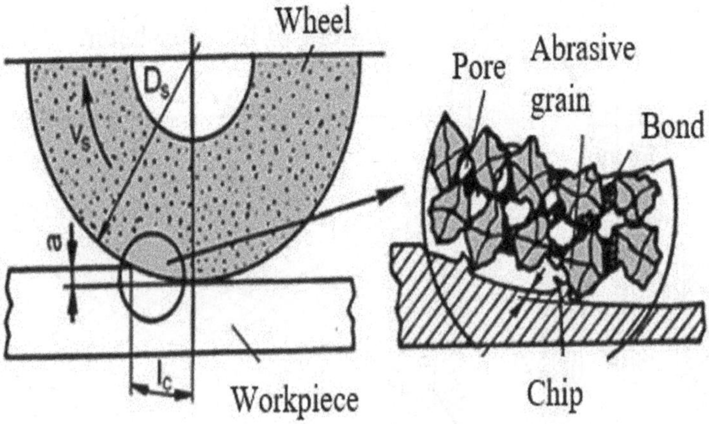

FIGURE 9.3 Photograph of the experimental setup of grinding process.

9.3 ADVANCED FINISHING PROCESSES WITHOUT EXTERNALLY CONTROLLED FORCES

9.3.1 Abrasive Flow Finishing (AFF) Process

AFF process is an advanced finishing process which was basically invented for deburring and finishing the cylindrical internal intricate surfaces (Rhoades, 1988). Figure 9.4 shows the setup for abrasive flow finishing and its components schematically.

AFF use the abrasive particles which are mixed in the carrier fluid for finishing the internal cylindrical surfaces. Carrier fluid having abrasives mixed in it is allowed to flow through internal surfaces of cylindrical workpieces under an extrusion pressure. Under the influence of extrusion pressure, abrasives erode surface roughness crests in the form of tiny chips, producing cylindrical internal finishing. The abrasives' abrasive effect in the constricting tube changes due to the rheological quality of the fluid (Loveless et al., 1994; Genc & Phule, 2002). The fluid's viscosity plays a crucial part in the finishing action (Jha & Jain, 2006). Chip removal is caused by the axial force acting on the abrasive particles, thus abrasive indentation into the workpiece's surface is caused by the radial force. The method is also used to complete non-ferromagnetic aircraft hydraulic and fuel system part (Lam & Smith, 1997; Jain, 2008). Components requiring uniform and repeatable outputs need this type of finishing processes. It can polish any surface through which fluid can flow. In this process, the abrading forces are not controllable during the finishing operation.

9.3.2 Elastic Emission Machining (EEM) Process

As can be observed in Figure 9.5, the EEM is a specific finishing process in which tiny abrasives target individual atoms on the surface of the workpiece, therefore, separating the material from the surface. The method is capable of extracting material from the workpiece's surface on an atomic scale and producing a smooth surface (Kanaoka et al., 2007;

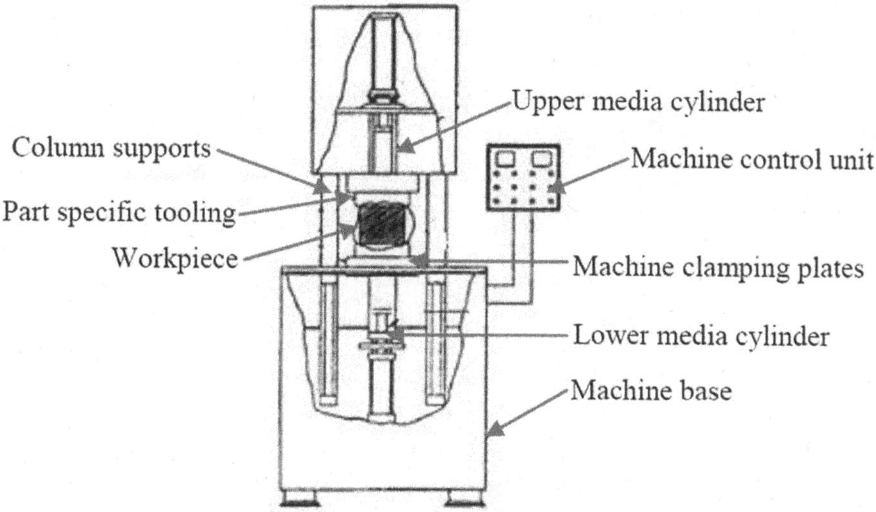

FIGURE 9.4 Schematic of abrasive flow finishing setup and its components.

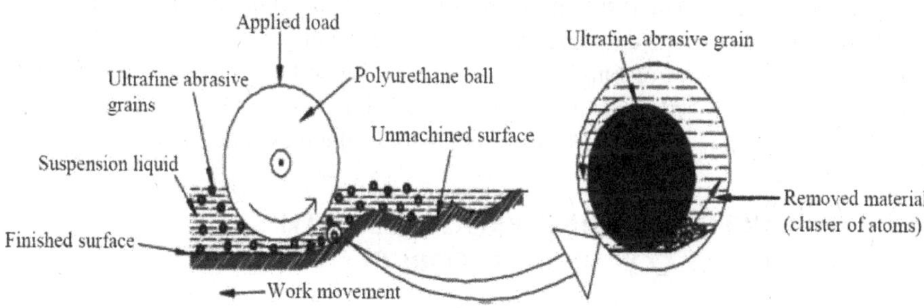

FIGURE 9.5 Elastic emission machining process.

Mori et al., 1990). It is a material-removal procedure in which the surface energy phenomenon is used to remove atoms due to abrasive interaction with the workpiece's surface. In this process, the inclusion of atomic scale fracture elastically produces super finished surface without any plastic deformation (Yamauchi et al., 2002). In this process also, the abrading forces are not controllable during the surface finishing.

9.3.3 Chemo-mechanical (CMP) Process

The CMP is an advanced finishing method in which the silica slurry and the workpiece undergo a chemical reaction (Bassea & Liangb, 1999; Komanduri et al., 1997; Seok, 2009). The process results in a surface that is free from surface defects. It's a finishing process that blends chemical and mechanical behavior, and it's useful in the electronics industry. The polishing rate is high, averaging 200 nm/min (however, this varies with the length of time spent polishing). The workpiece is polished under an applied load over the rotating polishing pad as depicted in Figure 9.6.

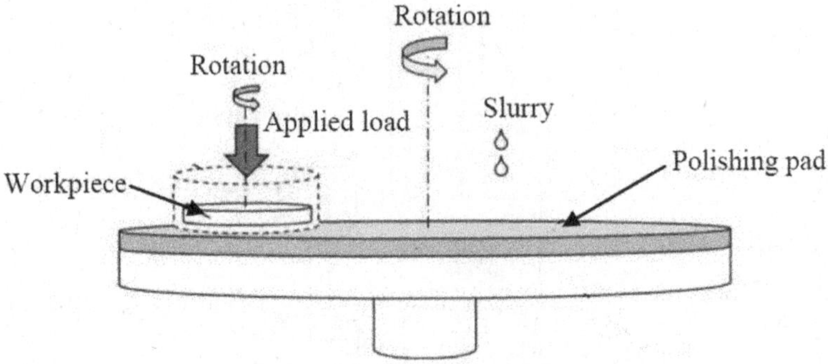

FIGURE 9.6 Process flow diagram for chemo-mechanical polishing.

The process is useful in semiconductor and electronics industries for polishing of silicon and optical and glass disc. As a surface finishing method, it avoids the issues of abrasive damage (such as scratching, brittle fracture, and surface pitting). For the time being, this technique can only be used to complete the finishing of flat surfaces. The finishing forces in the processes described earlier cannot be externally controlled, limiting their usefulness. As a result, these high-tech methods of finishing can cause flaws like micro cracks, surface damage, and even deeper levels of surface and sub-surface wear. Several new ways of finishing have been developed to overcome the problems with the aforementioned procedures, and they all share the ability to externally regulate the finishing forces by adjusting the intensity of the magnetic field.

9.4 ADVANCED FINISHING PROCESSES WITH EXTERNALLY CONTROLLED FORCES

By altering the current passing through the electromagnet coil, the magnetic field intensity in the working zone may be altered. As a result, the normal indenting force produced by the magnetic particles on the work-surface part is altered. Here are a few examples of sophisticated finishing techniques that are externally controlled.

9.4.1 Magnetic Abrasive Finishing (MAF) Process

The MAF is a sophisticated operation which uses either joined or unbonded particles in a magnetic field that functions like a flexible, self-deforming magnetic abrasive brush (FMAB) (Shinmura et al., 1990). The research was carried forward by Japan (Yamaguchi & Shinmura, 2004). The process was utilized to fine finish the inner or external surfaces of several applications (Umehara et al., 2006; Sidpara et al., 2009; Jayswal et al., 2005). Vibratory motion is performed in MAF to finish the edge and surface (Yamaguchi et al., 2007). Unbonded magnetic abrasive was used in the experiment on the cylindrical workpiece.

The ferromagnetic and SiC abrasives are combined with an SAE30 lubricant to create the unbonded magnetic abrasive. The cylindrical workpiece was subjected to the finishing examination using steel and iron grit ferromagnetic particles. Following testing, it has been determined that steel grit ferromagnetic particles, with their higher hardness and polyhedron shape, is better suitable for finishing the external surface of cylindrical workpieces (Umehara et al., 2006).

9.4.2 Static Flexible Magnetic Abrasive Brush (S-FMAB) Process

The static flexible magnetic abrasive brush (S-FMAB) process has limitation of non-uniform strength of abrasive brush. The reason for non-uniform strength is non-homogenous mixing of abrasive particles and iron particles. The surface roughness was reduced from 0.25 μm to 0.05 μm with the use of 180 μm steel grit mixed with 1.2 μm SiC abrasives. Figure 9.7 shows the initial and final MR finished surface images of the scanning electron microscopy and mirror, respectively. In S-FMAB, the current supplied is DC to the electromagnet and the chain of carbonyl iron particles (CIPs) with the abrasive particles is formed to perform the surface finishing. After finishing on some places, the workpiece has high roughness peaks at some places and has low roughness peaks at other places. The reason behind that is the non-homogeneous mixing of abrasives as well as CIPs. Limitation of S-FMAB was overcome by supplying the pulsating current to the electromagnet. This pulsating current allows the mixing of CIPs and abrasive particles in regular interval which improves the surface finishing obtained as compared to the S-FMAB (Chang et al., 2002).

In order to reduce tool wear, magnetic abrasive finishing is tested on the bare carbide surface (MAF). The uncoated carbide tool's flank, rake, and nose surfaces were used in the experiments. Each surface, such as the flank, rake, and nose, required a unique magnetic circuit configuration in order to ensure an even dispersion of magnetic particles. The tool's nose, rack, and flank were all finished on a five-axis, high-speed machining center. Experiments were run for 20 minutes on the flank surface, 15 minutes on the nose surface, and 20 minutes on the rake surface. The experimental results showed that Ra values on the flank and nose surfaces were less than 25 nm, while those on the rack surface were less than 50 nm. As a result of the finishing process, the authors found that the blades suffered minimal damage. As a bonus, this enhances tribological characteristics and lowers friction (Singh et al., 2005; Judal et al., 2014; Saraeian et al., 2016; Saraswathamma et al., 2015; Wang & Lee, 2009). A unique vibration-assisted cylindrical magnetic abrasive finishing system has been developed to polish cylinders. This study investigates the influence of the magnetic flux density, workpiece rotation speed, vibration frequency, and abrasive particle size on the smoothing of a cylinder's surface. The examination of the process variables revealed that the final smoothness of the cylinder's exterior is significantly influenced by the frequency of vibration and the size of the abrasive particles. In about 300 seconds, the roughness of the surface goes from 0.46 m to 0.18 mw (Kansal et al., 2018).

9.4.3 Magnetic Float Polishing (MFP) Process

Another method that may be adjusted to achieve ultrafine finishing of spheres such as ball bearings and ceramic balls by means of tiny abrasive particles is the MFP (Saraswathamma et al., 2015; Judal et al., 2013a; Sidpara & Jain, 2011; Jiao et al.,

2013; Seok et al., 2009; Kordonski & Jacobs, 1999; Kordonski & Golini, 1999; Jha & Jain, 2004, 2006; Jain et al., 1999; Michalski & Pawlus, 1992). The non-magnetic abrasives floating in the magnetic fluid experience levitation as a result of the fluid's hydrodynamic characteristics. The small abrasives and ferromagnetic particles that make up the magnetic polishing solution float in a foundation of kerosene oil or water. It is standard practice to utilize controlled finishing pressures in conjunction with magnetic float polishing (MFP) or magnetic abrasive finishing (MAF) for bringing the surface of exceptionally hard materials up to snuff. Stainless steel, non-ferrous metals, and ceramics are some examples of materials that benefit from these finishing procedures. Since the finishing medium in the MAF is often a dry powder made of iron and abrasives, its flow-ability with the brush on the surface of the work-piece is severely restricted. So MAF process with the dry magnetic brush may not be used for soft kind of materials. With the use of this process for soft materials, the problems such as scratches, sub-surface damage, or micro cracks may pertain over the surface. As a response to these issues, a wide variety of sophisticated finishing techniques using magnetorheological (MR) fluid, a smart fluid, are being developed.

9.5 MAGNETORHEOLOGICAL FLUID (MRF)

MRF is the intelligent fluid of choice for a wide range of manufacturing applications. Because its rheological characteristics may be adjusted rapidly, often within milliseconds, to meet specific needs, it has been dubbed a "smart fluid". In most cases, it functions when a magnetic field is present. To control the MR polishing fluid's rheological characteristics, engineers can adjust the magnetic flux density in the working zone. Under the dominance of the applied magnetic field, MR fluid parameters like yield strength and apparent viscosity change extremely quickly. MR fluid has a low shear strength and low viscosity, making it behave similarly to liquid when no magnetic field is present. In contrast, when the magnetic field is dominant, the fluid transforms into a viscous one with a high shear strength.

Figure 9.7 depicted that when an external magnetic field is not applied, the magnetic particles in MR fluid float freely in the base medium. When MR fluid is placed in a magnetic field, the magnetic particles in the fluid form chains along the field lines (as seen in Figure 9.7 [b]). MR fluids are made suitable for finishing by the addition of abrasive particles. MR polishing fluid, which contains abrasives, may be used to put the finishing touches on many different materials. When subjected to an external magnetic

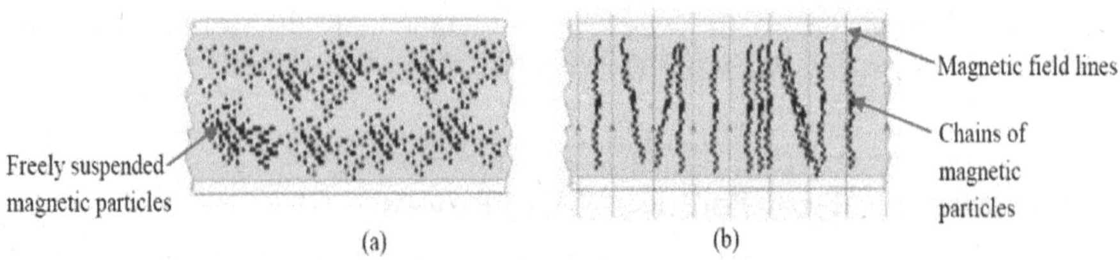

FIGURE 9.7 MR fluid (a) in the nonappearance of flux density and (b) in the existence of flux density.

field, magnetic iron particles arrange themselves into chains, creating a tight grip on the abrasives. Viscosity of MR fluid is altered by an external magnetic field (Kordonski & Jacobs, 1996; Bossis et al., 2002; Jolly et al., 1999; Judal et al., 2013a). The MR polishing fluid acts like a Bingham plastic fluid when subjected to an external magnetic field. In Figure 9.8, the rheological behavior of the magnetorheological fluid modeled by the Bingham plastic. When an external magnetic field is introduced, iron chains can trap more abrasive particles. The externally supplied magnetic field also regulates the MR polishing fluid's viscosity, making the fluid either more or less rigid (Kordonski & Golini, 1999; Kim et al., 1997). Magnetic iron particles also exert a levitational force on the abrasives, causing them to fly toward the direction of the workpiece's surface, where the magnetic field gradient is lowest. Because of the influence of the magnetic field, MR polishing fluid is resistant to sedimentation (Ashtiani et al., 2015). MR polishing fluid's smart properties of varying shear strengths make it more flexible for use in surface-finishing applications involving a wide range of workpiece materials (Vicente et al., 2011; Premalatha et al., 2012; Singh et al., 2011). The primary components of MR fluid are magnetic particles, which have a size on the order of microns. To construct the chains and secure the abrasives between them, the magnetic particles react to the applied magnetic field by aligning along the magnetic field lines. There are two main types of magnetic particles used in MR polishing fluid, carbonyl iron particles (CIPs) and electrolytic iron particles (EIPs). When finishing with MR fluids, magnetic iron particles are drawn to the tool's surface (a region with a high magnetic flux density), where they use levitation force to push back the abrasives on the workpiece's surface. Shearing action is employed to complete the task when non-magnetic abrasive particles are combined with an MR polishing solution (Carlson & Jolly, 2000; Haeri & Hashemabadi, 2008; Zantye et al., 2004; Wang et al., 2015b; Golini et al., 1999; Shorey et al., 2001; Lambropoulo et al., 1996; Sujith et al., 2019; Tu et al., 2019).

There are various kinds of abrasive available which are used in MR polishing fluid such as CeO2 for fine polishing of optical glasses, Al2O3 for the materials having low hardness, and diamond abrasives for high strength materials (Kordonski et al.,

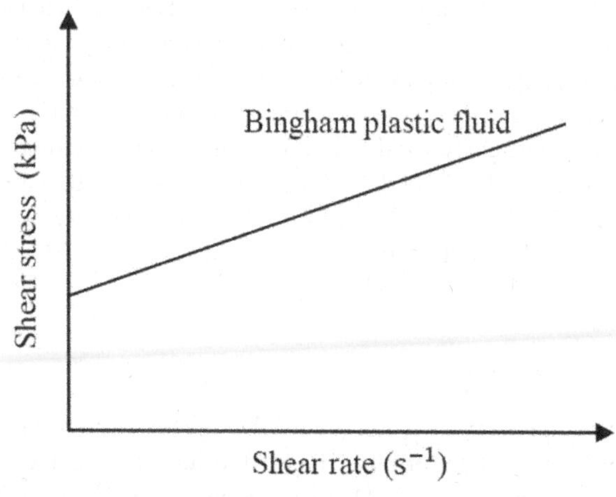

FIGURE 9.8 Bingham plastic model of magnetorheological fluid.

2006, 2007; Tricard et al., 2006; Jha & Jain, 2004; Furst & Gast, 2000; Das et al., 2011, 2010; Sadiq & Shunmugam, 2009; Singh et al., 2015, 2012a, 2012b, 2012c; Chen et al., 2016, 2017). The silicon carbide (SiC) abrasives are commonly used for finishing of metals. The kind of surface finishing required also depends on the shape and size of abrasives (Wang et al., 2015a; Singh et al., 2017, 2013; Brecker et al., 1969; Grover & Singh, 2017). Workpieces having high initial surface roughness are finished with abrasives of bigger size, whereas abrasives of lower size are utilized for fine surface finishing (Grover & Singh, 2018; Mann et al., 2016; Singh et al., 2016). In order to keep particles from settling to the bottom of the MR polishing fluid, certain additives are combined with the carrier fluids. Particles' consistent dispersion throughout the fluid is also maintained with the help of the additives. There are various types of additives available which are used for the synthesis of MR polishing fluid, such as grease, glycerol, or oleic acid. Some special additives like polystyrene and gaur gum are coated over the magnetic particle to prevent their oxidation in water-based carrier medium. There are varieties of carrier fluids available, such as oil, synthetic hydrocarbon, esters, water and paraffin oil heavy, etc. Carrier fluids like paraffin oil (heavy) are used in MR polishing for better flowability during the surface finishing of the metal workpieces (Singh et al., 2018, 2021; Singh & Singh, 2019, 2020a, 2020b, 2020c; Kumar et al., 2022). These fluids are being selected on the basis of various physical, chemical, or thermal (boiling temperature) properties according to the requirement of the finishing processes. Carrier fluid must be nonreactive with magnetic particles. Advanced finishing processes improves the functionality and life span of the industrial applications (Pathri et al., 2022; Babbar et al., 2022a, 2022b, 2022c, 2022d, 2021a, 2021b, 2021c, 2021d, 2021e, 2021f, 2020a, 2020b, 2020c, 2020d, 2020e, 2020f, 2019a; Sharma & Rai, 2022; Sharma et al., 2021, 2020a, 2020b; Khanduja et al., 2021; Prakash et al., 2021; Singh et al., 2021b, 2020). These processes are utilized for both the ductile and brittle material for fine finishing (Baraiya et al., 2020c; Babbar et al., 2019b, 2019c, 2019d, 2017; Kumar et al., 2019; Sharma et al., 2022a, 2022b, 2022c, 2021, 2019a, 2019b, 2018a, 2018b, 2021a; Sharma & Jain, 2020; Kumar et al., 2018).

Further, the potential of additive manufacturing (AM) to produce parts with intricate geometries has garnered a lot of attention in many studies. Laser polishing and other forms of post-production processing have been utilized to refine the surface and the result. The impacts of post-processing procedures on surface finish, stress levels, and fatigue performance are discussed, along with the current state of the art in these areas (Babbar et al., 2020, 2020a, 2020b, 2020c, 2020d; Kalia et al., 2022; Kumar et al., 2021; Parikh et al., 2023; Prakash et al., 2021; Rampal et al., 2021; Rana & Akhai, 2022; Sharma et al., 2018b, 2021a, 2021b, 2020a, 2020b, 2023a, 2023b, 2023c; Singh et al., 2021a, 2022a, 2023a).

9.6 CONCLUSION

Numerous technical applications employ the MRF technique to achieve the greatest finishing and overall product uniformity. The MRF is now the smart technology of the future as a result of its enhanced technology. External cylindrical finishing procedures are essential in the packaging sector. These procedures fulfill the specifications for cylindrical component finishing, improving machinery dependability and the automotive sector. In the mechanical industry, these processes are in high demand. Nano-level finishing is used to enhance the geometrical and functional features of cylindrical components in a number of applications.

REFERENCES

Ashtiani, M., Hashemabadi, S.H., & Ghaffari, A. (2015). A review on the magnetorheological fluid preparation and stabilization. *Journal of Magnetism and Magnetic Materials*, 374, 716–730.

Babbar, A., Jain, V., & Gupta, D. (2019c). Neurosurgical bone grinding. In *Biomanufacturing* (pp. 137–155). Cham: Springer International Publishing.

Babbar, A., Jain, V., & Gupta, D. (2019d). Thermogenesis mitigation using ultrasonic actuation during bone grinding: A hybrid approach using CEM43°C and Arrhenius model. *Journal of the Brazilian Society of Mechanical Sciences and Engineering*, 41.

Babbar, A., Jain, V., & Gupta, D. (2020e). In vivo evaluation of machining forces, torque, and bone quality during skull bone grinding. *Proceedings of the Institution of Mechanical Engineers, Part H: Journal of Engineering in Medicine*, 234, 626–638.

Babbar, A., Jain, V., Gupta, D., & Agrawal, D. (2021a). Histological evaluation of thermal damage to Osteocytes: A comparative study of conventional and ultrasonic-assisted bone grinding. *Medical Engineering & Physics*, 90, 1–8.

Babbar, A., Jain, V., Gupta, D., & Agrawal, D. (2021c). Finite element simulation and integration of CEM43 °C and Arrhenius Models for ultrasonic-assisted skull bone grinding: A thermal dose model. *Medical Engineering and Physics*, 90, 9–22.

Babbar, A., Jain, V., Gupta, D., & Prakash, C. (2021e). Experimental investigation and parametric optimization of neurosurgical bone grinding under bio-mimic environment. *Surface Review and Letters*, 2141005.

Babbar, A., Jain, V., Gupta, D., Prakash, C., & Agrawal, D. (2022c). Potential application of CEM43 °C and Arrhenius model in neurosurgical bone grinding. In *Numerical Modelling and Optimization in Advanced Manufacturing Processes* (pp. 145–158). Cham: Springer International Publishing.

Babbar, A., Jain, V., Gupta, D., Prakash, C., Agrawal, D., & Singh, S. (2021b). Experimental analysis of wear and multi-shape burr loading during neurosurgical bone grinding. *Journal of Materials Research and Technology*, 12, 15–28.

Babbar, A., Jain, V., Gupta, D., Prakash, C., & Sharma, A. (2020a). Fabrication and machining methods of composites for aerospace applications. In *Characterization, Testing, Measurement, and Metrology* (1st ed., pp. 109–124). Boca Raton, FL: CRC Press.

Babbar, A., Jain, V., Gupta, D., Prakash, C., Singh, S., & Pruncu, C. (2020f). Biomaterials and fabrication methods of scaffolds for tissue engineering applications. In *3D Printing in Biomedical Engineering* (pp. 167–186). Singapore: Springer.

Babbar, A., Jain, V., Gupta, D., Prakash, C., Singh, S., & Sharma, A. (2020b). Effect of process parameters on cutting forces and osteonecrosis for orthopedic bone drilling applications. In *Characterization, Testing, Measurement, and Metrology* (1st ed., pp. 93–108). Boca Raton, FL: CRC Press.

Babbar, A., Jain, V., Gupta, D., Prakash, C., Singh, S., & Sharma, A. (2020d). 3D bioprinting in pharmaceuticals, medicine, and tissue engineering applications. In *Advanced Manufacturing and Processing Technology* (1st ed., pp. 147–161). Boca Raton, FL: CRC Press.

Babbar, A., Jain, V., Gupta, D., & Sharma, A. (2020c). Fabrication of microchannels using conventional and hybrid machining processes. In *Non-Conventional Hybrid Machining Processes* (1st ed., pp. 37–51). Boca Raton, FL: CRC Press.

Babbar, A., Jain, V., Gupta, D., Sharma, A., Prakash, C., & Kumar, V. (2022b). Additive manufacturing for the development of biological implants, scaffolds, and prosthetics. In *Additive Manufacturing Processes in Biomedical Engineering* (pp. 27–46). Boca Raton, FL: CRC Press, Taylor Francis.

Babbar, A., Jain, V., Sharma, A., & Jain, A.K. (2019b). Rotary ultrasonic milling of C/SiC composites fabricated using chemical vapor infiltration and needling technique. *Materials Research Express*, 6, 085607.

Babbar, A., Kumar, A., Jain, V., & Gupta, D. (2019a). Enhancement of activated tungsten inert gas (A-TIG) welding using multi-component TiO2-SiO2-Al2O3 hybrid flux. *Measurement*, 148, 106912.

Babbar, A., Rai, A., & Sharma, A. (2021d). Latest trend in building construction: Three-dimensional printing. *Journal of Physics: Conference Series*, 1950, 012007.

Babbar, A., Sharma, A., Jain, V., & Gupta, D. (2022a). *Additive Manufacturing Processes in Biomedical Engineering: Advanced Fabrication Methods and Rapid Tooling Techniques*. Boca Raton, FL: CRC Press, Taylor Francis.

Babbar, A., Sharma, A., Kumar, R., Pundir, P., & Dhiman, V. (2021f). Functionalized biomaterials for 3D printing: An overview of the literature. In *Additive Manufacturing with Functionalized Nanomaterials* (pp. 87–107). Amsterdam: Elsevier.

Babbar, A., Sharma, A., & Singh, P. (2022d). Multi-objective optimization of magnetic abrasive finishing using grey relational analysis. *Materials Today: Proceedings*, 50, 570–575.

Babbar, A., Singh, P., & Farwaha, H.S. (2017). Parametric study of magnetic abrasive finishing of UNS C26000 flat brass plate. *International Journal of Advanced Mechatronics and Robotics*, 9, 83–89.

Balogun, V.A., & Mativenga, P.T. (2017). Specific energy based characterization of surface integrity in mechanical machining. *Procedia Manufacturing*, 7, 290–296.

Baraiya, R., Babbar, A., Jain, V., & Gupta, D. (2020c). In-situ simultaneous surface finishing using abrasive flow machining via novel fixture. *Journal of Manufacturing Processes*, 50, 266–278.

Benardos, P.G., & Vosniakos, G.C. (2003). Predicting surface roughness in machining: A review. *International Journal of Machine Tools and Manufacture*, 43, 833–844.

Beyerer, J., & Puente Leon, F. (1997). Detection of defects in groove textures of honed surfaces. *International Journal of Machine Tools and Manufacture*, 37(3), 371–389.

Bossis, G., Lacis, S., Meunier, A., & Volkova, A. (2002). Magnetorheological fluids. *Journal of Magnetism and Magnetic Materials*, 252, 224–228.

Brecker, J.N., Brown, R., Matsuo, T., Saito, K., Sweeney, J.A., Vansaun, J.B., & Shaw, M.C. (1969). Abrasive grain association on investigation of abrasive grain characteristics. *4th Annual Report*. Carnegie Institute of Technology, Pittsburgh, PA.

Carlson, J.D., & Jolly, M.R. (2000). Fluid, foam and elastomer devices. *Mechatronics*, 10, 555–569.

Chang, G.W., Yan, B.H., & Hsu, R.T. (2002). Study on cylindrical magnetic abrasive finishing using unbonded magnetic abrasives. *International Journal of Machine Tools and Manufacture*, 42(5), 575–583.

Chang, Y.P., Hashimura, M., & Dornfeld, D.A. (2000). An investigation of material removal mechanisms in lapping with grain size transition. *Journal of Manufacturing Science and Engineering*, 22(3), 413–419.

Chen, M., Liu, H., Su, Y., Yu, B., & Fang, Z. (2016). Design and fabrication of a novel magnetorheological finishing process for small concave surfaces using small ball-end permanent-magnet polishing head. *The International Journal of Advanced Manufacturing Technology*, 83(5–8), 823–834.

Chen, M., Liu, H., Su, Y., Yu, B., & Fang, Z. (2017). Model of the material removal function and an experimental study on a magnetorheological finishing process using a small ball-end permanent-magnet polishing head. *Applied Optics*, 56(19), 5573–5582.

Das, M., Jain, V.K., & Ghoshdastidar, P.S. (2010). Nano-finishing of stainless-steel tubes using rotational magnetorheological abrasive flow finishing process. *Machining Science and Technology*, 14(3), 365–389.

Das, M., Jain, V.K., & Ghoshdastidar, P.S. (2011). The out-of-roundness of the internal surfaces of stainless steel tubes finished by the rotational—magnetorheological abrasive flow finishing process. *Materials and Manufacturing Processes*, 26, 1073–1084.

Dimkovski, Z., Anderberg, C., Rosen, B.G., Ohlsson, R., & Thomas, T.R. (2009). Quantification of the cold worked material inside the deep honing grooves on cylinder liner surfaces and its effect on wear. *Wear*, 267(12), 2235–2242.

Furst, E.M., Gast, A.P. (2000). Micromechanics of magnetorheological suspensions. *Physics Revolution E*, 61(6), 6732–6739.

Genc, S., & Phule, P.P. (2002). Rheological properties of magnetorheological fluids. *Smart Materials and Structures*, 11, 140–146.

Golini, D., Kordonski, W.I., Dumas, P., & Hogan, S. (1999). Magnetorheological finishing (MRF) in commercial precision optics manufacturing. *Proceeding of SPIE Conference on Optical Manufacturing and Testing*, 3782, 80–91.

Grover, V., & Singh, A.K. (2017). A novel magnetorheological honing process for nano-finishing of variable cylindrical internal surfaces. *Materials and Manufacturing Processes*, 32(5), 573–580.

Grover, V., & Singh, A.K. (2018). Improved magnetorheological honing process for nano-finishing of variable cylindrical internal surfaces. *Materials and Manufacturing Processes*, 33(11), 1177–1187.

Gupte, P.S., Wang, Y., Miller, W., Barber, G.C., Yao, C. Zhou, B., & Zou, Q. (2008). A study of torn and folded metal (TFM) on honed cylinder bore surfaces. *Tribology Transactions*, 51, 784–789.

Haeri, S., & Hashemabadi, S.H. (2008). Three dimensional CFD simulation and experimental study of power law fluid spreading on inclined plates. *International Communications Heat and Mass Transfer*, 35, 1041–1047.

Jain, R.K., Jain, V.K., & Dixit, P.M. (1999). Modeling of material removal and surface roughness in abrasive flow machining process. *International Journal of Machine Tools and Manufacture*, 39, 1903–1923.

Jain, V.K. (2008). Abrasive-based nano-finishing techniques: An overview. *Machining Science and Technology*, 12, 257–294.

Jain, V.K. (2009). Magnetic field assisted abrasive based micro-nanofinishing. *Journal of Materials Processing Technology*, 209, 6022–6038.

Jayswal, S.C., Jain, S.C., & Dixit, P.M. (2005). Modeling and simulation of magnetic abrasive finishing process. *International Journal of Advanced Manufacturing Technology*, 26(5–6), 477–490.

Jha, S., & Jain, V.K. (2004). Design and development of magnetorheological abrasive flow finishing process. *International Journal of Machine Tool and Manufacture*, 44(10), 1019–1029.

Jha, S., & Jain, V.K. (2005). Nano-finishing techniques. In *Manufacturing and Nano Technology*, Editor: N.P. Mahalik (pp. 171–195). Heidelberg: Springer Verlag.

Jha, S., & Jain, V.K. (2006). Modeling and simulation of surface roughness in magnetorheological abrasive flow finishing (MRAFF) process. *Wear*, 261, 856–866.

Jiao, L., Wub, Y., Wang, X., Guo, H., & Liang, Z. (2013). Fundamental performance of magnetic compound fluid (MCF) wheel in ultra-fine surface finishing of optical glass. *International Journal of Machine Tools & Manufacture*, 75, 109–118.

Jolly, M.R., Bender, J.W., & Carlson, J.D. (1999). Properties and applications of commercial magnetorheological fluids. *Journal of Intelligent Material Systems and Structures*, 10(1), 5–13.

Judal, K.B., Yadava, V., & Pathak, D. (2013a). Experimental investigation of vibration assisted cylindrical magnet abrasive finishing of aluminium workpiece. *Materials and Manufacturing Processes*, 28, 1196–1202.

Judal, K.B., Yadava, V., & Pathak, D. (2013b). Study of vibration frequency and abrasive particle size during cylindrical Magnetic Abrasive Finishing. *International Journal of Precision Technology*, 3(2), 117–130.

Judal, K.B., Yadava, V., & Pathak, D. (2014). Experimental investigation of vibration assisted cylindrical-magnetic abrasive finishing of aluminum workpiece. *Materials and Manufacturing Processes*, 28, 1196–1202.

Kalia, G., Sharma, A., & Babbar, A. (2022). Use of three-dimensional printing techniques for developing biodegradable applications: A review investigation. *Materials Today: Proceedings*, 62, 346–352.

Kanaoka, M., Liu, C., Nomura, K., Ando, M., Takino, H., & Fukuda, Y. (2007). Figuring and smoothing capabilities of elastic emission machining for low-thermal-expansion glass optics. *Journal of Vacuum Science & Technology B, Nanotechnology and Microelectronics: Materials, Processing, Measurement, and Phenomena*, 25, 2110–2113.

Kansal, H., Singh, A.K., & Grover, V. (2018). Magnetorheological nano-finishing of diamagnetic material using permanent magnets tool. *Precision Engineering*, 51, 30–39.

Khanduja, P., Bhargave, H., Babbar, A., Pundir, P., & Sharma, A. (2021). Development of two-dimensional plotter using programmable logic controller and human machine interface. *Journal of Physics: Conference Series*, 1950, 012012.

Khanna, O.P., & Lal, M. (2010). *A Text Book of Production Technology*. New Delhi, India: Dhanpat Rai Publication.

Kim, J.D., Kang, Y.H., Bae, Y.H., & Lee, S.W. (1997). Development of a magnetic abrasive jet machining system for precision internal polishing of circular tubes. *Journal of Material Processing Technology*, 71, 384–393.

Komanduri, R., Lucca, D.A., & Tani, Y. (1997). Technological advances in fine abrasive processes. *CIRP Annals—Manufacturing Technology*, 46(2), 545–596.

Kordonski, W.I., & Golini, D. (1999). Fundamentals of magnetorheological fluid utilization in high precision finishing, *Journal of Intelligent Material Systems and Structures*, 10(9), 683–689.

Kordonski, W.I., & Jacobs, S.D. (1996). Magnetorheological finishing. *International Journal of Modern Physics B*, 10, 2837–2848.

Kordonski, W.I., & Jacobs, S.D. (1999). Progress update in magnetorheological finishing. *International Journal of Modern Physics B*, 13, 2205–2212.

Kordonski, W.I., & Shorey, A.B. (2007). Magnetorheological (MR) jet finishing technology. *Journal of Intelligent Material Systems and Structures*, 18, 1127–1130.

Kordonski, W.I., Shorey, A.B., & Tricard, M. (2006). Magnetorheological (MR) jet finishing technology. *Journal of Fluids Engineering*, 128(1), 20–26.

Kumar, M., Babbar, A., Sharma, A., & Shahi, A.S. (2019). Effect of post weld thermal aging (PWTA) sensitization on micro-hardness and corrosion behavior of AISI 304 weld joints. *Journal of Physics: Conference Series*, 1240, 012078.

Kumar, M., Sharma, A., & Shahi, A.S. (2018.) *A Sensitization Studies on the Metallurgical and Corrosion Behavior of AISI 304 SS Welds*, Vol. PartF7. Singapore: Springer.

Kumar, S., Sudhakar, R.P., Goyal, D., & Sehgal, S. (2021). Process modelling for machining Inconel 825 using cryogenically treated carbide insert. *Metal Powder Report*, 76, 66–74. https://doi.org/10.1016/j.mprp.2020.06.001

Kumar, V., Prakash, C., Babbar, A., Choudhary, S., Sharma, A., & Uppal, A.S. (2022). *Additive Manufacturing Processes in Biomedical Engineering* (pp. 143–164). Boca Raton, FL: CRC Press, Taylor Francis.

Lam, S.S.Y., & Smith, A.E. (1997). Process monitoring of abrasive flow machining using a neural network predictive model. *Industrial Engineering Research-Conference Proceedings* (pp. 477–482). https://www.eng.auburn.edu/~aesmith/files/saraierc.pdf

Lambropoulo, S.J., Yang, F., & Jacob, S.D. (1996). Optical fabrication and testing. *Technical Digest Series (Optical Society of America, Washington DC)*, 7, 150–153.

Larsen-Bassea, J., & Liangb, H. (1999). Probable role of abrasion in chemo-mechanical polishing of tungsten. *Wear*, 233–235, 647–654.

Lawrence, K.D., & Ramamoorthy, B. (2011). An accurate and robust method for the honing angle evaluation of cylinder liner surface using machine vision. *International Journal of Advanced Manufacturing Technology*, 55, 611–621.

Loveless, T.R., Williams, R.E., & Rajurkar, K.P. (1994). A study of the effects of abrasive-flow finishing on various machined surfaces. *Journal of Materials Processing Technology*, 47, 133–151.

Mann, S., Singh, G., & Singh, A.K. (2016). Nano surface finishing of permanent mould punch using MR fluid based finishing processes. *Materials and Manufacturing Processes*, 32(9), 1004–1010.

Michalski, J., & Pawlus, P. (1992). Description of honed cylinders surface topography. *International Journal of Machine Tools and Manufacture*, 34(2), 199–210.

Mohanty, H.K., Mahapatra, M.M., Kumar, P., Biswas, P., & Mandal, N.R. (2013). Predicting the effects of tool geometries on friction stirred aluminium welds using artificial neural network and fuzzy logic techniques. *International Journal of Manufacturing Research*, 8(3), 296–312.

Mori, Y., Yamauchi, K., Endo, K., Ide, T., Toyota, H., Nishizawa, K., & Hasegawa, M. (1990). Evaluation of elastic emission machined surface by scanning tunnelling microscopy. *Journal of Vacuum Science & Technology A: Vacuum, Surfaces, and Films*, 8, 621–624.

Parikh, P., Sharma, A., Trivedi, R., Roy, D., & Joshi, K. (2023). Performance evaluation of an indigenously-designed high-performance dynamic feeding robotic structure using advanced additive manufacturing technology, machine learning and robot kinematics. *International Journal on Interactive Design and Manufacturing (IJIDeM)*, 1–29.

Pathri, B.P., Khan, M.S., & Babbar, A. (2022). Relevance of bio-inks for 3D bioprinting. In *Additive Manufacturing Processes in Biomedical Engineering* (pp. 81–98). New York, USA: Taylor & Francis.

Pattnaik, S., Karunakar, D.B., & Jha, P.K. (2012). Developments in investment casting process—A review. *Elsevier Journal of Materials Processing Technology*, 212, 2332–2348.

Pawlus, P., Cieslak, T., & Mathia, T. (2009). The study of cylinder liner plateau honing process. *Journal of Materials Processing Technology*, 20, 6078–6086.

Prakash, C., Kumar, V., Mistri, A., Sharma, A., Uppal, A.S., Babbar, A., & Pathri, B.P. (2021). Investigation of functionally graded adherents on failure of socket joint of FRP composite tubes. *Materials*, 14, 6365.

Premalatha, S.E., Chokkalingam, R., & Mahendran, M. (2012). Magneto mechanical properties of iron based MR fluids. *American Journal of Polymer Science*, 2(4), 50–55.

Rajurkar, K.P., Levy, G., Malshe, A., Sundaram, M.M., McGeough, J., Hu, X., Resnick, R., & Desilva, A. (2006). Micro and nano machining by electro-physical and chemical processes. *Annals of the CIRP*, 55, 643–666.

Rampal, R., Goyal, T., Goyal, D., Mittal, M., Dang, R.K., & Bahl, S. (2021). Magnetorheological abrasive finishing (MAF) of soft material using abrasives. *Materials Today: Proceedings*, 45, 51140–5121. https://doi.org/10.1016/j.matpr.2021.01.629

Rana, M., & Akhai, S. (2022). Multi-objective optimization of Abrasive water jet Machining parameters for Inconel 625 alloy using TGRA. *Materials Today: Proceedings*, 65, 3205–3210. https://doi.org/10.1016/j.matpr.2022.05.374

Rhoades, L.J. (1988). Abrasive flow machining. *Manufacturing Engineering*, 1, 75–78.

Sabri, L., & Mansori, M.E. (2009). Process variability in honing of cylinder liner with vitrified bonded diamond tools. *Surface and Coatings Technology*, 204, 1046–1050.

Sabri, L., Mezghani, S., Mansori, M.E., & Zahouani, H. (2011). Multiscale study of finish-honing process in mass production of cylinder liner. *Wear*, 27, 1509–1513.

Sadiq, A., & Shunmugam, M.S. (2009). Investigation into magnetorheological abrasive honing (MRAH). *International Journal of Machine Tools and Manufacture*, 49, 554–560.

Saraeian, P., Mehr, H.S., Moradi, B., Tavakoli, H., & Alrahmani, O.K. (2016). Study of magnetic abrasive finishing for AISI321 stainless steel. *Materials and Manufacturing Processes*, 31(15), 2023–2029.

Saraswathamma, K., Jha, S., & Rao, P.V. (2015). Experimental investigation into ball end magnetorheological finishing of silicon. *Precision Engineering*, 42, 218–223.

Seok, J., Lee, S.O., Jang, K.I., Min, B.K., & Lee, S.J. (2009). Tribological properties of a magnetorheological (MR) fluid in a finishing process. *Tribology Transaction*, 52(4), 460–469.

Shaji, S., & Radhakrishnan, V. (2003). Analysis of process parameters in surface grinding with graphite as lubricant based on the Taguchi method. *Journal of Materials Processing Technology*, 141, 51–59.

Sharma, A., Babbar, A., Jain, V., & Gupta, D. (2018b). Enhancement of surface roughness for brittle material during rotary ultrasonic machining. *MATEC Web of Conferences*, 249, 01006.

Sharma, A., Babbar, A., Tian, Y., Pathri, B.P., Gupta, M., & Singh, R. (2022a). Machining of ceramic materials: A state-of-the-art review. *International Journal on Interactive Design and Manufacturing (IJIDeM)*, 1–21. https://doi.org/10.1007/s12008-022-01016-7

Sharma, A., Fidan, I., Huseynov, O., Ali, M.A., Alkunte, S., Rajeshirke, M., Gupta, A., Hasanov, S., Tantawi, K., Yasa, E., Yilmaz, O., Loy, J., & Popov, V. (2023b). Recent inventions in additive manufacturing: Holistic review. *Inventions*, 8(4), 103. https://doi.org/10.3390/inventions8040103

Sharma, A., Grover, V., Babbar, A., & Rani, R. (2020b). A trending nonconventional hybrid finishing/machining process. In *Non-Conventional Hybrid Machining Processes* (1st ed., pp. 79–93). Boca Raton, FL: CRC Press.

Sharma, A., & Jain, V. (2020). Experimental investigation of cutting temperature during drilling of float glass specimen. In *IOP Conference Series: Materials Science and Engineering* (Vol. 715, No. 1, p. 012050). IOP Publishing. https://doi.org/10.1088/1757-899X/715/1/012050

Sharma, A., Jain, V., & Gupta, D. (2018a). Characterization of chipping and tool wear during drilling of float glass using rotary ultrasonic machining. *Measurement*, 128, 254–263.

Sharma, A., Jain, V., & Gupta, D. (2018c). *Tool Wear Analysis While Creating Blind Holes on Float Glass Using Conventional Drilling: A Multi-Shaped Tools Study* (Vol. PartF7). Singapore: Springer.

Sharma, A., Jain, V., & Gupta, D. (2019a). A novel investigation study on float glass hole surface integrity & tool wear using Chemical assisted Rotary ultrasonic machining. *Materials Today: Proceedings*, 26, 632–637.

Sharma, A., Jain, V., & Gupta, D. (2019b). Multi-shaped tool wear study during rotary ultrasonic drilling and conventional drilling for amorphous solid. *Proceedings of the Institution of Mechanical Engineers, Part E: Journal of Process Mechanical Engineering*, 233, 551–560.

Sharma, A., Jain, V., & Gupta, D. (2019c). Comparative analysis of chipping mechanics of float glass during rotary ultrasonic drilling and conventional drilling: For multi-shaped tools. *Machining Science and Technology*, 23, 547–568.

Sharma, A., Jain, V., & Gupta, D. (2021a). Effect of pre and post tempering on hole quality of float glass specimen: For rotary ultrasonic and conventional drilling. *Silicon*, 13, 2029–2039.

Sharma, A., Jain, V., & Gupta, D. (2022b). Mathematical approach on chipping volume estimation generated during rotary ultrasonic drilling for float glass. *Proceedings of the National Academy of Sciences, India Section A: Physical Sciences*, 92, 285–291.

Sharma, A., Jain, V., Gupta, D., & Babbar, A. (2020a). A review study on miniaturization. In *Advanced Manufacturing and Processing Technology* (1st ed., pp. 111–131). Boca Raton, FL: CRC Press.

Sharma, A., Kalsia, M., Uppal, A.S., Babbar, A., & Dhawan, V. (2022c). Machining of hard and brittle materials: A comprehensive review. *Materials Today: Proceedings*, 50, 1048–1052.

Sharma, A., Kumar, V., Babbar, A., Dhawan, V., & Kotecha, K. (2021b). Experimental investigation and optimization of electric discharge machining process parameters using grey-fuzzy-based hybrid techniques. *Materials*, 14(19), 5820. https://doi.org/10.3390/ma14195820

Sharma, A., & Rai, A. (2022). Fused deposition modelling (FDM) based 3D & 4D printing : A state of art review. *Materials Today: Proceedings*, 62, 367–372. https://www.sciencedirect.com/science/article/abs/pii/S2214785322020582

Sharma, A., Sandhu, H.S., Goyal, D., Goyal, T., Jarial, S., & Sharda, A. (2023a). Sustainable development in cold gas dynamic spray coating process for biomedical applications: Challenges and future perspective review. *International Journal on Interactive Design and Manufacturing (IJIDeM)*, 1–17. https://doi.org/10.1007/s12008-023-01474-7

Sharma, V.K., Rana, M., Singh, T., Singh, A.K., & Chattopadhyay, K. (2021). Multi-response optimization of process parameters using desirability function analysis during machining of EN31 steel under different machining environments. *Materials Today: Proceedings*, 44, 3121–3126. https://doi.org/10.1016/j.matpr.2021.02.809

Shinmura, T., Takazawa, K., Hatano, E., & Matsunaga, M. (1990). Study on magnetic abrasive finishing. *Annals of CIRP*, 39(1), 325–328.

Shorey, A.B., Jacobs, S.D., Kordonski, W.I., & Gans, R.F. (2001). Experiments and observations regarding mechanism of glass removal in magnetorheological finishing. *Applied Optics*, 40(1), 20–33.

Sidpara, A., Das, M., & Jain, V.K. (2009). Rheological characterization of magnetorheological finishing fluid. *Materials and Manufacturing Processes*, 24(2), 1467–1478.

Sidpara, A., & Jain, V.K. (2011). Experimental investigations into forces during magnetorheological fluid based finishing process. *International Journal of Machine Tools and Manufacture*, 51, 358–362.

Singh, A.K., Jha, S., & Pandey, P.M. (2011). Design and development of nanofinishing process for 3D surfaces using ball end MR finishing tool. *International Journal of Machine Tools and Manufacture*, 51, 142–151.

Singh, A.K., Jha, S., & Pandey, P.M. (2012a). Magnetorheological ball end finishing process. *Materials and Manufacturing Processes*, 27(4), 389–394.

Singh, A.K., Jha, S., & Pandey, P.M. (2012b). Nanofinishing of a typical 3D ferromagnetic workpiece using ball end magnetorheological finishing process. *International Journal of Machine Tools and Manufacture*, 63, 21–31.

Singh, A.K., Jha, S., & Pandey, P.M. (2012c). Nanofinishing of fused silica glass using ball-end magnetorheological finishing tool. *Materials and Manufacturing Processes*, 27(10), 1139–1144.

Singh, A.K., Jha, S., & Pandey, P.M. (2013). Mechanism of material removal in ball end magnetorheological finishing process. *Wear*, 302, 1180–1191.

Singh, A.K., Jha, S., & Pandey, P.M. (2015). Performance analysis of ball end magnetorheological finishing process with MR polishing fluid. *Materials and Manufacturing Processes*, 30(12), 1482–1489.

Singh, B.P., Singh, J., Bhayana, M., & Goyal, D. (2021b). Experimental investigation of machining nimonic-80A alloy on wire EDM using response surface methodology. *Metal Powder Report*, 76, 9–17. https://doi.org/10.1016/j.mprp.2020.12.001

Singh, B.P., Singh, J., Bhayana, M., Singh, K., & Singh, R. (2022). Experimental examination of the machining characteristics of Nimonic 80-A alloy on wire EDM. *Materials Today: Proceedings*, 69, 291–296. https://doi.org/10.1016/j.matpr.2022.08.537

Singh, D.K., Jain, V.K., & Raghuram, V. (2005). On the performance analysis of flexible magnetic abrasive brush. *Machining Science and Technology*, 9(4), 601–619.

Singh, G.S., Babbar, A., Jain, V., & Gupta, D. (2021a). Comparative statement for diametric delamination in drilling of cortical bone with conventional and ultrasonic assisted drilling techniques. *Journal of Orthopaedics*, 25, 53–58.

Singh, G.S., Singh, A.K., & Garg, P. (2016). Development of magnetorheological finishing process for external cylindrical surfaces. *Materials and Manufacturing Processes*, 32(5), 581–588.

Singh, G.S., Singh, A.K., & Garg, P. (2017). Development of magnetorheological finishing process for external cylindrical surfaces. *Materials and Manufacturing Processes*, 32(5), 581–588.

Singh, J., Singh, C., & Singh, K. (2023). Rotary ultrasonic machining of advance materials: A review. *Materials Today: Proceedings*. https://doi.org/10.1016/j.matpr.2023.01.159

Singh, M., & Singh, A.K. (2019). Improved magnetorheological finishing process with rectangular core tip for external cylindrical surfaces. *Materials and Manufacturing Processes*, 34(9), 1049–1061.

Singh, M., & Singh, A.K. (2020a). Magnetorheological finishing of grooved drum surface and its performance analysis in winding process. *International Journal of Advanced Manufacturing Technology*, 106, 2921–2937.

Singh, M., & Singh, A.K. (2020b). Magnetorheological finishing of copper cylindrical roller for its improved performance in printing machine. *Proceedings of the Institution of Mechanical Engineers, Part E: Journal of Process Mechanical Engineering*, https://doi.org/10.1177/0954408920945232

Singh, M., & Singh, A.K. (2020c). Theoretical investigations into magnetorheological finishing of external cylindrical surfaces for improved performance. *Journal of Mechanical Engineering Science, Proceedings of the Institution of Mechanical Engineers Part C*, 234(24), 4872–4892.

Singh, M., Singh, A.K., & Singh, A.K. (2018). A rotating core-based magnetorheological nanofinishing process for external cylindrical surfaces. *Materials and Manufacturing Processes*, 33(11), 1160–1168.

Singh, S., Prakash, C., Pramanik, A., Basak, A., Shabadi, R., & Królczyk, G. (2020). Magneto-rheological fluid assisted abrasive nanofinishing of β-Phase Ti-Nb-Ta-Zr alloy: Parametric appraisal and corrosion analysis. *Materials*, 13, 5156.

Spencer, A., Almqvist, R., & Larsson, R. (2011). A numerical model to investigate the effect of honing angle on the hydrodynamic lubrication between a combustion engine piston ring and cylinder liner. *Proceedings of the Institution of Mechanical Engineers, Part J: Journal of Engineering Tribology*, 225(7), 683–689.

Sujith, S.V., Solanki, A.K., & Mulik, R.S. (2019). Experimental evaluation on rheological behavior of Al2O3-pure coconut oil nanofluids. *Journal of Molecular Liquids*, 286, 110905.

Tricard, M., Kordonski, W.I., Shorey, A.B., & Evans, C. (2006). Magnetorheological jet finishing of conformal, freeform and steep concave optics. *CIRP Annals—Manufacturing Technology*, 55(1), 309–312.

Tu, H.X., Thao, L.P., Hong, T.T., Nga, N.T.T., Trung, D.D., Gong, J., & Pi, V.N. (2019). Influence of dressing parameters on surface roughness of workpiece for grinding hardened 9XC tool steel. *IOP Conference Series: Material Science Engineering*, 542, 012008.

Umehara, N., Kirtane, T., Gerlick, R., Jain, V.K., & Komanduri, R. (2006). A new apparatus for finishing large size large batch silicon nitride (Si3N4) balls for hybrid bearing applications by magnetic float polishing (MFP). *International Journal of Machine Tools and Manufacture*, 46, 151–169.

Verma, G.C., Kala, P., & Pandey, P.M. (2017). Experimental investigations into internal magnetic abrasive finishing of pipes. *International Journal of Advanced Manufacturing Technology*, 88(5), 1657–1668.

Vicente, V.J., Klingenberg, D.J., & Hidalgo-Alvarez, R. (2011). Magnetorheological fluids: A review. *Soft Matter*, 7, 3701–3710.

Wang, A.C., & Lee, S.J. (2009). Study the characteristics of magnetic finishing with gel abrasive. *International Journal of Machine Tools and Manufactures*, 49, 1063–1069.

Wang, J., Chen, W., & Han, F. (2015a). Study on the magnetorheological finishing method for the WEDMed pierced die cavity. *International Journal of Advance Manufacturing Technology*, 76, 1969–1975.

Wang, Y.Q., Yin, S.H., Hunag, H., Chen, F.J., & Deng, G.J. (2015b). Magnetorheological polishing using a permanent magnetic yoke with straight air gap for ultra-smooth surface planarization. *Precision Engineering*, 40, 309–317.

Yamaguchi, H., & Shinmura, T. (2004). Internal finishing process for alumina ceramic components by a magnetic field assisted finishing process. *Precision Engineering*, 28, 135–142.

Yamaguchi, H., Shinmura, T., & Ikeda, R. (2007). Study of internal finishing of austenite stainless steel capillary tubes by magnetic abrasive finishing. *Journal of Manufacturing Science and Engineering*, 129, 885–892.

Yamauchi, K., Yamamura, K., Mimura, H., Sano, Y., Saito, A., Souvorov, A., Yabashi, M., Tamasaku, K., Ishikawa, T., & Mori, Y. (2002). Nearly diffraction- limited line focusing of 149 a hard x-ray beam with a elliptically figures mirror. *Journal of Synchrotron Radiation*, 9, 313–316.

Zantye, P.B., Kumar, A., & Sikder, A.K. (2004). Chemical mechanical planarization for microelectronics applications. *Material Science and Technology*, 45, 189–220.

Index

P

Q

R

S

T

U

V

W

X

Y

Z